POWER-UP

Unlocking the Hidden Mathematics in Video Games

MATTHEW LANE

PRINCETON UNIVERSITY PRESS
Princeton and Oxford

Published by Princeton University Press, 41 William Street,
Princeton, New Jersey 08540
In the United Kingdom: Princeton University Press, 6 Oxford Street,
Woodstock, Oxfordshire OX20 1TR

press.princeton.edu

Names: Lane, Matthew, 1982–
Title: Power-up : unlocking the hidden mathematics in video games / Matthew Lane.
Description: Princeton : Princeton University Press, [2017] | Includes
bibliographical references and index.
Identifiers: LCCN 2017011065 | ISBN 9780691161518 (hardcover)
Subjects: LCSH: Mathematics—Computer-assisted instruction. | Games in
mathematics education. | Video games.
Classification: LCC QA20.C65 L36 2017 | DDC 510-dc23 LC record available at
https://lccn.loc.gov/2017011065

British Library Cataloging-in-Publication Data is available

This book has been composed in Archer Book

Printed on acid-free paper. ∞

Typeset by Nova Techset Pvt Ltd, Bangalore, India
Printed in the United States of America

1 3 5 7 9 10 8 6 4 2

POWER-UP

For Linus, in hopes of preemptively
answering the question,
When will I ever need to use math?

Contents

Acknowledgments

First of all, thanks to you, dear reader, for checking out this book. And doubly thanks for checking out the acknowledgments! This demonstrates a degree of interest that bodes well for our time together, I hope.

At the risk of stating the obvious, writing a book is hard. A lot of people supported me through this process; without them, I would not have had the privilege of writing this acknowledgments section (to say nothing of the pages that follow it). First I'd like to thank Vickie Kearn at Princeton University Press for encouraging me to write this book, and for her unyielding enthusiasm about the project. Her assistant, Lauren Bucca, was also incredibly supportive in the final stretch of this process, and always patiently answered the myriad of questions I had.

Many of my questions pertained to securing image permissions for the book. This ended up being a somewhat involved process and required the help of a number of people. In no particular order, I'd like to thank KT Vick, Mia Putrino, Besty Ek, and Tom Dillof at Sony; Zack Cooper and Heather Pond at Ubisoft; Ferry Halim, creator of the Orisinal games; Frank Goldberg and Dan Garon at Zynga; Rik Eberhardt at MIT Game Lab; Casandra Brown at Atari; Gerilynn Maria at Blue Planet Software; and Gena Feist at Take-Two Interactive.

Before moving on from permissions, I'd like to send an extra special thank you to Marc ten Bosch and Asher Vollmer. Not only did they let me use screenshots from their games in this book, but they also agreed to my interview requests and allowed me to reprint portions of those interviews. Thanks for being so generous with your time.

While writing can often feel like a solitary pursuit, I wouldn't have been able to finish this book without the support of my friends and family. Thanks to the many people who connected me with people in the video game industry, as well as those who jumped at the chance to brainstorm titles. I'd also like to thank all of the friends with whom I've played games or talked about math over the years; those experiences have absolutely shaped the thinking that led to this book.

Finally, the biggest thanks of all is reserved for my family. Thanks to my parents and grandparents for fostering a love of learning early on and encouraging my curiosity wherever it led. And thank you a thousand times over to my wife, Meg, who has had to endure countless nights and weekends of me writing and rewriting. Thank you for your patience, unwavering support, and gentle reminders that things like eating and sleeping are basic human needs I shouldn't neglect. There's no way I could have done this without you.

But enough about me. Let's do some math!

Introduction

The late 1980s were an exciting time. The Nintendo Entertainment System (better known as the NES) was just a few years old in America, and *Super Mario Bros.* was already a hit. In the summer of 1987, Nintendo released two more gaming blockbusters: *The Legend of Zelda* and *Metroid.* In so doing, it quickly cemented itself as a leader in the nascent video game industry. Though I didn't yet own an NES, I quickly befriended classmates who did.

As in *Super Mario Bros.*, the plot of the original *Legend of Zelda* involved a guy in funny clothes running around in order to rescue a princess. But beyond that, the two games shared few similarities. In *Super Mario Bros.*, players progressed from left to right through a series of essentially linear stages. Players could easily measure their progress by watching the digits increase as they moved from Level 1-1 to Level 8-4. In contrast, *The Legend of Zelda* simply dropped players in the middle of a forest with little indication of where to go. Players were encouraged to explore a world in which progression was nonlinear, restricted only by the abilities of the main character, a precocious boy named Link. The world itself was open, but Link was initially not experienced enough to fully navigate it. To break through a cracked wall, he needed bombs; to cross a river, he needed a raft. As you became more familiar with the game, so too did Link discover the tools needed to navigate his surroundings.

A similar level design was used in *Metroid.* In this game, the player took control of an intergalactic bounty hunter named Samus. Her mission: run-and-gun her way through a colony of space pirates in order to stop them from creating biological weapons on the planet Zebes. Though the side-scrolling perspective made the game look like a sort of *Super Mario Bros.* in outer space, the soul of the game was much closer to that of *Zelda.* Instead of bombs, Samus used missiles; instead of a raft, she enhanced her space suit to access new areas. In both games, upgrading your character was part and parcel of exploring the world.

The Legend of Zelda series is now one of the most successful video game franchises in the world. Nintendo has produced more than fifteen games in the series, which collectively have sold tens of millions of copies worldwide. *Metroid*'s numbers, while not quite as large, are in the same ballpark. Clearly, these games resonate with players. But why? And what does any of this have to do with mathematics?

To address this question, let's jump ahead a few years to the very end of the twentieth century. In 1994, Princeton professor Andrew Wiles solved one of the most famous open problems in all of mathematics. Known as Fermat's Last Theorem, the problem asserts that the equation $x^n + y^n = z^n$ has no solutions when x, y, z, and n are whole numbers with n greater than 2. This is in sharp contrast to the case $n = 2$. In this case, examples of solutions to $x^2 + y^2 = z^2$ for integer x, y, and z are familiar to geometry students. Such solutions are referred to as Pythagorean triples, and there are infinitely many of them: $x = 3$, $y = 4$, $z = 5$ is one example ($3^2 + 4^2 = 5^2$), as is $x = 5$, $y = 12$, $z = 13$, ($5^2 + 12^2 = 13^2$).

Though the problem is simple enough to state, its solution eluded mathematicians for hundreds of years. Wiles' proof of the theorem (completed with the help of Richard Taylor) garnered him instant recognition, even outside of the mathematical community. Contributing to his newfound fame was a 1997 *NOVA* episode on the problem and Wiles' solution. The episode opens with Wiles describing on what it feels like to do mathematics:

> Perhaps I could best describe my experience of doing mathematics in terms of entering a dark mansion. One goes into the first room, and it's dark, completely dark. One stumbles around bumping into the furniture, and gradually, you learn where each piece of furniture is, and finally, after six months or so, you find the light switch. You turn it on, and suddenly, it's all illuminated. You can see exactly where you were.

The word choice here is particularly apt. If mathematics is a house, then certainly it must be a large one. The mansion of mathematics consists of a vast collection of rooms, and the light from one sometimes only barely illuminates the room across the hall. If your goal is to make a floor plan of the entire house, you've got quite a tall task ahead of you.

The longer you fumble around through the mansion, though, the clearer things will become. Some rooms will eventually become fully illuminated. If you're diligent, entire wings of the mansion will open up to you. You'll also begin to find connections between different areas of the house that would have been completely surprising to you in the dark. Maybe you discover that the small hallway off of the kitchen provides a more direct route to the sitting room. Perhaps you'll be able to return to a previously discovered locked door and unlock it from the other side.

While we should be careful about stretching this metaphor beyond its natural limits, it does provide a helpful way to think about how some mathematicians work. The sometimes inelegant initial search for truth; the building of connections between seemingly disjointed areas; the development of tools to aid in the solution to a problem—these are all hallmarks of the mathematician's quest for understanding.

But these are also hallmarks of the game design featured in *Zelda*, *Metroid*, and many other video game franchises. Without a map to guide the way or a linear world that made navigation a simple matter of moving from left to right, exploration of these early game worlds was often rife with missteps, wrong turns, and confusion about where to go or what to do. With enough patience and skill, however, the game world's internal logic began to show itself. A locked door on the planet Zebes might eventually reveal itself to be a shortcut between two formerly distant parts of the planet, but only after Samus obtained the artillery necessary to blast the door open. To reach an entirely new portion of the kingdom of Hyrule, Link might first need to get his hands on a ladder to help him climb up the ominously named Death Mountain. I don't mean to suggest that solving Fermat's Last Theorem is an achievement anywhere near comparable to finding a virtual upgrade that allows you to freeze space pirates with ice (although the latter may make for a more entertaining way to spend an afternoon). I do contend, however, that there are some similarities between doing mathematics and playing games. Moreover, I believe that exploring the intersection of mathematics and video games can help to build a deeper appreciation for both.

Beyond these high-level qualitative comparisons, there are a number of more concrete and quantitative connections between mathematics and video games. While sometimes surprising, these links arise organically enough once you've developed an eye for spotting them. Here's just a sampling of some of the questions we'll explore together:

- What do video games have to teach us about our physical world ... and more important, what aren't they teaching us?
- Why is it that certain classic games, such as *Jeopardy* and *Pictionary*, make for lackluster video games?
- What's the best way to measure the quality of user-generated content in video games, and what can this teach us about the mathematics of voting?
- How do those pesky red shells in *Mario Kart* work, and what's the best strategy for avoiding them?

Of course, there are many differences between doing mathematics and playing video games. For example, no matter how many hours you spend in the mansion of mathematics, there will always remain rooms cloaked in darkness. The worlds presented in many video games, however, are finite. This is why people can spend their entire career doing mathematics without ever getting bored, while many video games, once finished, sit idly on a shelf. But by looking at a wide assortment of games across time and different platforms, the mansion of video games grows substantially. While it still may not be as large as the mansion of mathematics, between these two domiciles there's more than enough to inspire some fun and interesting questions.

Although we'll be talking about math and video games, this is not a book about game theory. It's also not necessarily about game development, even though there's some juicy math there. And while tablets and Chromebooks have given developers new avenues to explore games for the classroom, I'm not necessarily interested in games whose purpose is to explicitly help students learn mathematics. Mostly this is because, with some exceptions, most educational math games aren't great at being educational or great at being games.

Instead, the primary goal of these chapters is to search for interesting and sometimes unexpected bits of mathematics using video games as a springboard. The moral here is that beautiful mathematics can be found in the unlikeliest of places. And by using video games as the entry point, my hope is to make mathematics topics at all levels more accessible to a general audience.

If you are a prolific gamer, I'd like to get you to think about the games you play a little differently, and encourage you to seek out mathematical ideas in those games. If you're a teacher, maybe something in these pages will give you a new idea for a way to inspire your students. If you are a lover of both games and mathematics, then welcome! You should feel right at home here.

If you love video games but hate math, here comes the bad news: we'll be doing some mathematics in this book. You should be able to overcome the mathematical barrier to entry, however, provided you've taken some high school math classes (think geometry, algebra, trigonometry, and the like). I'll do my best to explain what's going on under the mathematical hood in the simplest possible terms. If you're a more advanced reader, technical facts and discussions have generally been relegated to endnotes[1] and appendices. My aim is to reach as large an audience as possible, so if you're one of many people with an ingrained math phobia, please don't fret. There will be plenty here for you to enjoy.

Finally, as with writing about anything in popular culture, there is always a danger that this book will become irrelevant by the time it reaches the reader. Thankfully, beautiful mathematics never goes out of fashion, and so to combat this, one only needs to be careful in the selection of games. To that end, I've adopted the following guidelines when deciding on whether or not to discuss a certain game or franchise. Though it may not garner me much street cred with the hardcore gaming community, the examples I will discuss draw primarily from games that have already established themselves and are reasonably wellknown or are based on other established properties. This may be bad news for obscure titles but should be good news for you, since, with rare exception, I hope you will have heard of the games we'll discuss. While the latest Mario game may fade in popularity, Mario himself does not appear to be going anywhere.

Here's how the book is structured. In the first chapter, we'll take a look at realism (or lack thereof) in games and explore how a game's reality can affect learning. We'll also take a look at a few games that are often touted as exemplars when it comes to the intersection of commercial games and education.

In the chapters that follow, we'll examine games that don't typically come to mind when thinking about education and highlight examples of rich mathematics that can emerge from these games. The chapters are organized thematically—one focuses on the mathematics of voting, another on computational complexity, and so on—and for the most part these chapters are logically independent. Feel free to peruse them in any order you like.

Once we've explored a wide variety of examples, the last chapter focuses on a simple question: *why games*? After all, aren't games simply a diversion from the more important things we should actually be doing? To close things out, I'll try to convince you (if I haven't already) why games have value.

Though mathematics is undeniably one of humankind's noblest pursuits, for many people it simply isn't as engaging as killing zombies or building empires. But part of this is a marketing problem, one that mathematics has suffered from for far too long. What follows is my attempt to convince you that mathematics can be fun and frequently appears in unexpected places; or, if you need no convincing, to at least give you some interesting examples to motivate mathematical thinking.

Ready? Let's-a go!

1
Let's Get Physical

1.1 PLATFORMING PERILS

In one of my first high school physics classes, our teacher gave us a diagnostic exam to see what sort of misconceptions we had about the everyday world. The questions were fairly run of the mill as far as introductory physics goes: *Which will fall faster, a bowling ball or a golf ball? If you swing a ball on a string above your head, in which direction will it travel when you let go?*

By and large, intro physics students don't perform well on these assessments. But students don't always know this. Sometimes the questions seem so easy, or the misconceptions are so deeply ingrained, that the gap between perceived and actual performance can be quite large. And while I no longer have the results of my own diagnostic, I remember having to come to terms with this gap in my nascent (and now dead) physics career.

Rather than take responsibility for my own lack of understanding, I'd like to officially blame Nintendo instead. For without the success of games like *Super Mario Bros.*, perhaps I would have been better equipped to answer questions like this:

> If you go skydiving, what will your trajectory look like after you step out of the plane but before you open up your parachute?[1] (Ignore air resistance.)

See Figure 1.1 for three possible answers. The correct one is on the left. As soon as you leave the plane, the only force acting on you is gravity, which pushes you towards the Earth. If you ignore air resistance, the instant you step off the plane your horizontal velocity will be the same

FIGURE 1.1. The trajectory on the left represents what would actually happen to an object falling out of a plane (ignoring air resistance).

as the plane's. Because of this, relative to someone watching you on the ground, you will fall forward, not straight down or backward.[2] Just because you leave the plane doesn't mean you lose the velocity you accumulated while on it.

Having played more than my fair share of *Super Mario Bros.* as a child, this feature of our world was somewhat counterintuitive to me. Had Mario stepped out of a plane, I thought, he would have fallen straight down. I made this assessment based on countless afternoons spent running and jumping as Mario through the Mushroom Kingdom.

Consider what happens when Mario jumps while standing on a moving platform.[3] If you've ever accompanied Mario on one of his two-dimensional adventures, you may recall that when you press the jump button while Mario is standing still, he'll jump straight up. It doesn't matter if the platform is also standing still or is racing across the screen; once you hit that button, Mario jumps as though he had been standing on a stationary surface.

Let's get a little more specific. Suppose Mario is standing still on a platform that's moving slowly to the right. Since he's standing still, his velocity will be the same as the velocity of the platform. But as soon as you press the jump button, this inherited velocity disappears (Figure 1.2).

It makes no sense for Mario's horizontal motion to stop when he jumps. If this were how things worked in the real world, jumping up on an airplane would cause you to slam into the back of it. But this

FIGURE 1.2. Screen shots of a video game character defying physics. When he jumps up, he stops moving forward. Screenshots from *Super Mario Bros Video Game*. © Nintendo

isn't what happens: assuming the air is calm and you're at cruising altitude, jumping on an airplane zipping through the sky doesn't feel any different than jumping on the ground. In my first physics class, however, such a real-world example didn't come immediately to mind. Instead, I found myself thinking about the physics of Mario, rather than, you know, *actual* physics.

This is just one reason why students looking to *Super Mario Bros.* for physical intuition aren't likely to have much success. But this isn't even the most obvious reason: how about the fact that Mario's height doubles when he eats mushrooms? How is it possible that Mario can jump at least twice his height? How is he able to reverse the direction of his motion while in the *middle* of a jump?

In my defense, *Super Mario Bros.* wasn't the only game to throw the laws of physics out of the window. In fact, a whole genre of games helped to reinforce these types of misconceptions. This genre is aptly named the *platformer*, a blanket term used to describe any game in which jumping from platform to platform is an essential component of the gameplay. For a time in the 1980s and 1990s the platformer was the most popular game genre, and many video game icons have their roots in platforming. Unfortunately, these games often had no regard for realistic physics.

This isn't to say that every platform game of the late twentieth century treated the laws of physics so cavalierly. Indeed, a notable exception to the rule is the *Castlevania* series. Released in America in 1987, the first game was popular enough to spawn two sequels before the end of the decade. There are now more than thirty games in the series, which centers on a family of vampire hunters and their (apparently) never-ending quest to vanquish the undead.

In the first few games, the main character could not alter his trajectory mid-jump. Once you pressed that jump button, the character fully committed to jumping left, right, or straight up. This is much more in line with how our world actually works. No matter how athletic you might be, there's no way you could jump forward, change your mind in midair, turn around, and land back where you started.

Games like *Castlevania* were in the minority, however. And even though this particular series remains popular,[4] a vast majority of platforming games at the time played fast and loose with the physics of jumping. As a result, the mechanics in a game like *Castlevania*—one that strived for a more realistic approach—felt clunky and rigid in comparison. Ironically, a more realistic platformer felt *less* natural.

There's a rich history of inaccurate physics in games. In this chapter, we'll explore several examples of the intersection between our reality and video game reality, and examine some implications for learning. We'll also take a look at games that explore reality in ways that are unique to the medium of video games. Finally, we'll take a quick look at *Minecraft*, a game with little regard for realism but one that is often touted as a great educational resource. Throughout this chapter, we'll be taking a critical eye toward the medium, in order to give us a better understanding of the current state of video games and education.

1.2 PLATFORMING IN THREE DIMENSIONS

As video game technology has improved, so too has its depiction of reality. Sometimes this is out of necessity: as games become more realistic, disconnects between the physics of the game world and the physics of the real world become easier to notice. But even games that don't strive for realism sometimes adopt mechanics that make their self-contained worlds a little more authentic.

Take our good friend Mario, for instance. In the fall of 1996, he made his first foray into the third dimension with the release of *Super Mario 64*, the not-so-creatively-named flagship title for the Nintendo 64. The physics model in the game was still far from realistic: Mario could jump higher and farther than any man reasonably could, and for the most part he still had far too much control over his movement when he was airborne.

When it came to jumping while on a moving platform, however, this was the first game in the series to get it right. In *Super Mario 64*, if you pressed the jump button while Mario was standing on a moving platform, he maintained the platform's velocity while in the air. For the first time, you could have Mario jump straight up without necessarily worrying about him falling to his death. This small concession to the real world has remained in all of Mario's subsequent 3D adventures. In fact, more recent Mario games have placed the plumber in scenarios where physics—specifically, gravity—has been an important focus. In *Super Mario Galaxy* and its sequel, released for the Nintendo Wii in 2007 and 2010, respectively, many of the environments Mario explores are in outer space. Because of this, platforms may have their own gravitational fields, and navigating Mario through these environments introduces plenty of challenges. The orbits he makes as he is pulled in different directions by nearby celestial bodies can be difficult to predict. Perhaps if I had grown up a few years later, I would have performed slightly better on my first physics assessment.

Then again, perhaps not. Although some of the everyday effects of gravity may be more faithfully represented in more recent games, digging deeper reveals some more fundamental problems.

For example, in 2014, Claire Sullivan and some of her colleagues at the University of Leicester published a paper titled "It's a-me Density!" on a problem with some of the planets Mario visits on his interstellar adventures. Specifically, many of the planets are quite small—mere meteors, really—and yet, in spite of this, Mario jumps as high on these planets as he does back on Earth.

In order for these microplanets to have gravitational pulls as strong as their larger colleagues, this means the smaller planets must be dense. Really dense.

How dense, you ask? Well, according to the authors of the paper, so dense that most of these planets couldn't actually exist in real life. With gravitational pulls as strong as the game suggests, these planets would be forced—pun intended—to explode. So much for using Mario's three-dimensional adventures as a way to build intuition about our

FIGURE 1.3. Sackboy has an ah-mazing wardrobe. Screenshots from *LittleBigPlanet*. © Sony

physical reality. At least in this case, though, the misconception arises in a scenario that's fairly far removed from our everyday experiences.

This isn't to say that students can't learn anything about physics by playing platform games. Indeed, quite a few teachers use these games specifically to address misconceptions or highlight certain implausible features of the game world.[5] But without proper guidance, these games (regardless of the number of dimensions in which they're played) can also reinforce misconceptions rather than help dispel them. It's a lesson I learned all too well during my first week of high school physics.

1.3 *LITTLEBIGPLANET*: EXPLORING PHYSICS THROUGH GAMEPLAY

I'm living proof that there's a risk students will adopt misconceptions by playing certain games. But is there any way for teachers to use games to help students learn physics the right way? Many game developers hope the answer is yes.

One such developer is Media Molecule, the creative force behind the popular PlayStation franchise *LittleBigPlanet*.[6] The star of the franchise is Sackboy (Figure 1.3), who can run, jump, and swing his way through a variety of colorful and well-designed two-dimensional worlds. In this sense, the series is a bit of a throwback to the platforming games of yesteryear, as the genre as a whole has waned in popularity since its heyday in the '80s and '90s.

But when *LittleBigPlanet* was released for the PlayStation 3 in 2008, the critical acclaim it received wasn't simply due to 2D platforming nostalgia. What really set the game apart was its successful blending of classic platforming game mechanics with more modern social elements and robust customization tools.

For example, the levels are littered with items that allow players to customize Sackboy in millions of different ways. Some of these items are easy to find, but in order to find everything, players will need some creative problem-solving skills and the assistance of up to three of their friends.

But this barely scratches the surface of the customization options that are available. Each game in the *LittleBigPlanet* series also comes with a fully functional level editor so that players can create their own worlds for Sackboy to explore. Indeed, the original motto for the game was "Play, create, share"; this philosophy was fully embraced by players, who as of October 2014 had created and uploaded more than nine million levels.[7]

As you might expect, the level editor offers a variety of options for would-be game designers. You can make platforms out of different materials. You can adorn your level with any number of stickers and decorations. You can even select a general visual theme for your level and choose the music to play during different sections.

All of this is secondary, though, to the literal nuts and bolts of *LittleBigPlanet*'s level editor. The game is built on a relatively realistic physics engine (Sackboy maintains velocity when jumping from a moving platform, but he has more control over his in-air trajectory than would a real-life sack person), and many of the tools in the level creation tool kit work precisely because the physics is so believable. These tools include, but are not limited to, bolts, glue, string, rods, springs, pistons, winches, two-way switches, three-way switches, and explosives. Taken together, these tools form a rich playground of possibilities for any student of mechanics.

People have used these tools to create all kinds of crazy levels. Some, like elaborate calculating machines,[8] are more interesting from a mathematical perspective than a gameplay one. But the level editor also provides some opportunities for students to explore concepts in physics.

The folks at Media Molecule have tried to tap into this potential, with mixed results. Their first attempt was a series of "Contraption

Challenges"—events in which players competed against one another to build specific objects using the game's level editor. Challenges ranged from building a flying machine to a Rube Goldberg device.[9] To date, there have been more than a dozen challenges, and while the prizes for the winners have typically been restricted to virtual accessories like crowns and pins, the prestige of winning is at least somewhat real.[10]

For those of us in need of more practical incentives, there are people willing to pay serious money to explore the question of whether or not games can help us teach math and science. For example, every year since 2007 the MacArthur Foundation[11] has teamed up with the Humanities, Arts, Science, and Technology Alliance and Collaboratory (HASTAC for short) to sponsor a Digital Media and Learning Competition. During its third year, and in conjunction with Sony Computer Entertainment America, the competition featured a category called "Game Changers." Contestants were tasked with using *LittleBigPlanet* to create content that helped students learn science and math. Cash prizes for the competition ranged from $7,500 to $40,000.

In the end, five competitors were awarded prizes for their submissions. There are levels about math, physics, chemistry, and even stem cell technology, but unfortunately none of them leverage *LittleBigPlanet* to create something truly engaging.[12] The stem cell level, for instance, is primarily a text-driven experience—go here, read this, then go there, read that—and some of the levels appear to have been published before they were finished. When it comes to learning, most of these levels are clunky or incomplete.

This doesn't necessarily mean that games can't help students learn physics. It could be that the *LittleBigPlanet* series has yet to realize its educational potential, or that *LittleBigPlanet* itself isn't the best vehicle for exploring topics in math and physics. Either way, it's debatable whether or not the series currently has a lot of solutions for education. However, there are other games that might offer up a bit more.

1.4 FROM 2D TO 3D: BENDING LAWS IN *PORTAL*

When Mario made the leap into the third dimension, the physics of his universe inched ever so slightly closer to reality. And indeed, as technology has improved and developers have created more and more immersive three-dimensional worlds, the physics engines within those worlds have improved as well. Perhaps, then, if we want to use a game to explore the physics around us in depth, we should focus our attention on 3D games.

Modern game libraries feature plenty of 3D titles from which to choose. Nevertheless, there is one series of 3D games that clearly stands above the rest when it comes to integrating interesting physics. Like *LittleBigPlanet*, it has also already dabbled in education. The series is called *Portal*, and quite simply, it is wonderful.

To the casual observer, *Portal* might look like just another first-person shooter. After all, the game is viewed from the perspective of a gun-toting character. But unlike many other first-person shooters, the gun in *Portal* is not used to harm. In fact, aside from a few laser-eyed robots, the game is fairly nonviolent.

Instead of shooting bullets, the gun in *Portal* is used to shoot, well, portals. The entire premise of the game is built around this portal gun, which serves as a kind of teleportation device. Fire a beam in one spot, then another, and a portal is created that connects these two locations. The player can walk through the portal to instantaneously travel between these two spots.

As you can imagine, this opens the game up to a lot of interesting possibilities. Many of the puzzles involve figuring out how to position your portals to gain access to places that would be otherwise impossible to reach. But it's about much more than finding the right spots; because the portals conserve momentum, many puzzles require an understanding of the game's physics in order to proceed.

Here's an example. Suppose you enter a room that has its only exit on the far side of a long chasm. The chasm is too far to jump across, and any attempt to make the leap will only leave you stranded at the bottom of the pit.

Variations on this type of puzzle appear throughout the *Portal* series. The main idea is to open a portal with one end at the bottom of the pit and one end high up on the wall by the entrance. When you jump into the portal at the bottom of the chasm, gravity will accelerate you, and your momentum will propel you through the portal and across the pit (Figure 1.4).

This type of problem solving is ripe for exploration in the classroom. Valve, the developer of *Portal* and its sequel, clearly realized this. Just

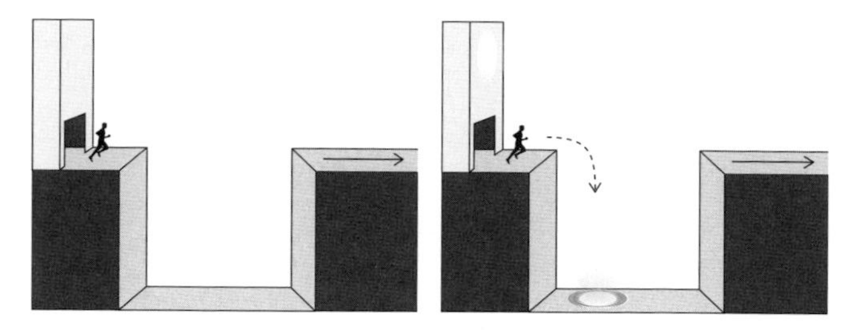

FIGURE 1.4. One possible *Portal* puzzle. By creating a portal at the bottom of the gap (blue), and one at top of the wall (orange), the player can use momentum to try and shoot across the room.

over a year after the release of *Portal 2* in the spring of 2011, Valve unveiled the *Teach with Portals* initiative, aimed at integrating science and math education with the *Portal* series. Through *Teach with Portals,* educators are able to access *Portal 2* as well as a level editor, so that they can create their own levels within the game. The *Teach with Portals* website also serves as a sort of hub where teachers who have used the level editor to create lessons can share their resources. There are a number of fairly detailed lesson plans on the site, with topics ranging from oscillatory motion to terminal velocity.

Many of the lessons rely, either implicitly or explicitly, on the fact that the portals conserve momentum. In other words, if you throw something (including a person) through one end of the portal, that something will exit the portal with the same velocity.[13] It doesn't take long, though, to realize that if the portals conserve momentum, then they must also be doing something weird with energy, a quantity which must also be conserved.[14]

For example, suppose you place an object on the ground so that it sits motionless. The energy of the system will be zero.[15] Now suppose you put one portal on the floor directly below your object. No biggie; the energy of the object is still zero.

But now, let's suppose you put a second portal above the object. What will happen?

To answer this question, take a look at Figure 1.5. The object will fall through the portal below it and exit through the portal on the ceiling. Gravity will then force the object down to the ground and through the portal once more. The object will fall through the portal over and

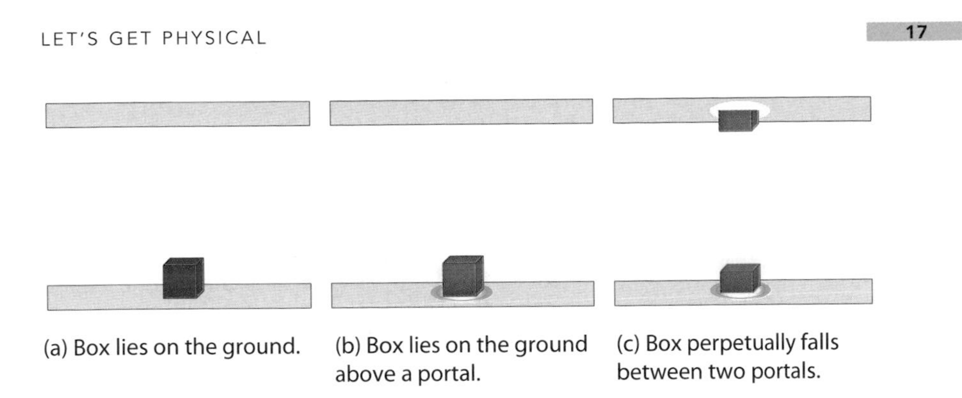

FIGURE 1.5. Creating a perpetual motion machine with portals.

over, accelerating until the upward force of air resistance balances the downward force of gravity. When this happens, its velocity stays the same, and we say that the object has reached its *terminal velocity*. Upon reaching its terminal velocity it will fall at a constant speed from portal to portal until the player moves one of the portals or turns off the game.

It would seem that these portals have effectively created a perpetual motion machine. While this may sound awesome, perpetual motion machines can't exist in the real world, and so, unfortunately, neither can these portal guns.[16] Of course, the lack of energy conservation is also an issue. Notice that the object falling through the portal goes from having zero energy to a nonzero amount of energy; moreover, it will be dissipating energy to the atmosphere in the form of heat and sound, even after it reaches its terminal velocity.

And we haven't even specified the object traveling through the portal! While there may be size restrictions on allowable objects (something wider than the portal won't fit through, obviously), there are no mass restrictions. A heavy object will fall through this setup just as easily as a light one, possibly picking up even more energy before it reaches its terminal velocity. Putting this all together, it seems like in order to comply with the law of conservation of energy, the portals must be drawing from a seemingly endless energy supply. Since the portals essentially work autonomously once your gun is fired, how this is possible remains quite a mystery.

Does this mean that the *Portal* series can't work as a suitable educational resource? Not necessarily. Talking about the reasonableness of the model in these games is certainly valuable, and even ignoring these paradoxes, there are still interesting questions that can be asked without bumping up against some of the thornier issues. Nevertheless, since portals give rise to physical contradictions, using the *Portal* series as a teaching tool may be risky if students form misconceptions while playing the game. But in the hands of an experienced teacher, many pitfalls can be anticipated, and using these games to enhance the curriculum may be worth considering.[17]

1.5 EXPLORING REALITY WITH *A SLOWER SPEED OF LIGHT*

Conversations about whether hypothetical teleportation portals violate this law or that law can be difficult to resolve, because the entire premise is fictional and there are limits to what can be verified and tested. Moreover, it doesn't take much intellectual prodding for the physical model in the *Portal* series to begin to show cracks, and if the model itself violates physical principals, some might argue that there's little point in even using the game to study physics. After all, what's the point in rigorously analyzing a broken system?

At the same time, there's something to be said for the way in which games like *Portal* subvert our everyday experiences about the way the world works. Even though *Portal*'s game mechanics (by which I mean the set of rules governing our interactions with the game) may not hold up to strict scrutiny, within the game world there's a relatively consistent internal logic that's extremely compelling and still affords teachers the ability to ask interesting questions.

Having said that, wouldn't it be even better if we could use games that maintained some of the counterintuitive gameplay features in *Portal* but were built on a rock-solid physical foundation? Games that would allow for deep and rich exploration outside of our everyday experience but that would also more accurately reflect the natural laws of our universe? Developers are beginning to grapple with this question as well and are coming up with some unique gameplay experiences.

As you might expect, one group that spends a lot of time thinking about these questions is the MIT Game Lab. Originally founded as the Singapore-MIT GAMBIT Game Lab in 2006, the MIT Game Lab officially opened its doors in the fall of 2012 and shortly thereafter released its first game, *A Slower Speed of Light*. The game attempts to do what our real-world experiences simply can't: make Einstein's theory of special relativity seem more intuitive.

If you're looking for a book on relativity, you've come to the wrong place. But, so that you can understand the purpose of the game, here's a brief summary of the theory's salient features. Everything rests on the following two assumptions:

1. The laws of physics are the same in any two reference frames when one frame moves at a constant velocity with respect to the other frame.
2. The speed of light is a universal constant, i.e., it does not vary with the reference frame.

Let's parse this a bit. The simplest way to do so is with some examples. The first assumption says, for instance, that if you are practicing your juggling skills in your backyard, with some careful thinking you'll be able to derive Newton's laws of motion, formulate laws for conservation of energy and momentum, and so on. Meanwhile, if I do the same thing while on a speeding train, I'll find the exact same physical laws that you found—provided the train is moving with a constant velocity.

What's more, absent any external factors (peeking outside the window, checking the GPS on my phone,[18] etc.), I won't know whether my train car is moving or at rest. As long as the velocity is constant, my frame of reference will give no indication as to the value of its velocity. In particular, if I throw one of my juggling balls straight up in the air, it will land right back in my hand, rather than being throttled back toward the rear of the train. Objects within a reference frame that moves at constant velocity maintain that velocity. To put it another way, the physical model in a game like *Super Mario Bros.* is broken on a fundamental level, as it fails to conform to the first postulate of relativity!

Constant velocity is a key piece of the puzzle here. If the train I'm on undergoes some change in acceleration—if the conductor slams on the breaks, for example, or we pull into a station and the train's speed decreases to zero—then of course I will be able to detect this type of motion (I may feel a force pushing me away from my seat, for example). But with no acceleration, my world will feel as though it is at rest.

Provided our frames of reference are related to each other by a constant velocity, it seems that nature does not prefer one of us to the other when it comes to uncovering its laws.

Next, let's turn to the second assumption. The significance of this assumption may not seem at all clear. This is due in large part to the fact that our everyday experiences happen at speeds much slower than the speed of light (roughly 671 million miles per hour).

To combat this pedagogically inconvenient reality, let's imagine a universe in which all the laws of physics as we know them still hold, but the speed of light is much lower. In such a world, we might find ourselves moving closer to the speed of light on an everyday basis, and the consequences of the second postulate of special relativity would be more apparent.

This thought experiment lies at the heart of *A Slower Speed of Light.*[19] In the game, your character collects objects that lower the speed of light. The farther you progress in the game, the slower the speed of light becomes. When every object has been collected (not a difficult feat; the game can be completed in well under ten minutes), you are given time to roam around the game area and observe the effects of relativity in all their glory. Though the effects are best experienced by playing the game,[20] the screenshots in Figure 1.6 illustrate one of the strangest consequences of the second postulate: the warping of space for objects moving near the speed of light.

This effect, known as *length contraction,* is a direct consequence of the postulates of relativity. But because it is so contrary to our everyday experience, it took time for length contraction to gain wide acceptance as a property of our world.[21] (For an explanation of *why* length contraction occurs, see the addendum to this chapter.)

Even though the effects of relativity are now accepted as true, length contraction still is not something that is easy to visualize except in the simplest of cases. By basing a game mechanic on this effect, the designers at MIT Game Lab are doing with games what can't be done in any other medium: offering a fully immersive experience that allows us to explore otherwise inaccessible features of our universe.

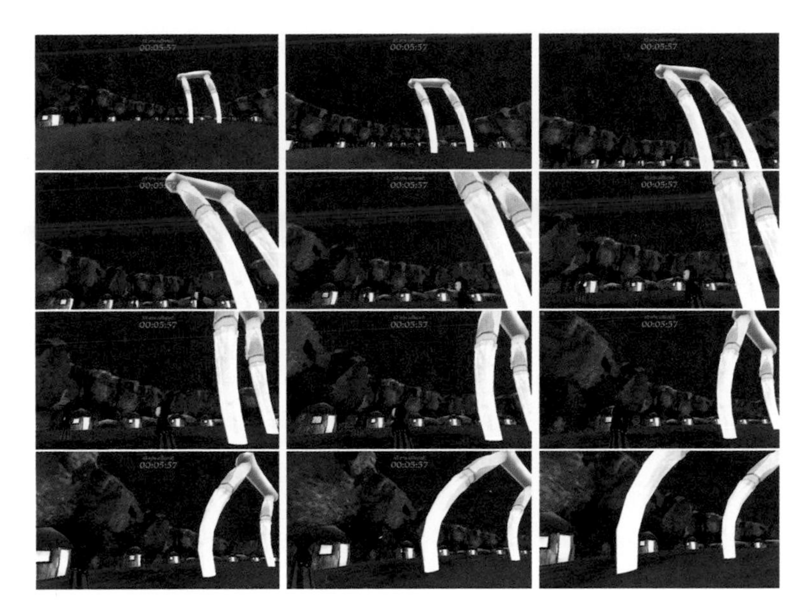

FIGURE 1.6. Exploring length contraction in *A Slower Speed of Light*. Although everything gets distorted as you move, the tall arch in the center of town shows the most pronounced effects. Reproduced with permission from MIT Game Lab.

1.6 EXPLORING ALTERNATIVE REALITIES

A Slower Speed of Light explores a real phenomenon that is totally foreign to our everyday experience. But video games can do more than highlight the extremes of our reality. They can also provide an opportunity to explore meaningful alternatives to our reality.

For example, by now you're probably quite familiar with your everyday existence in a world with three dimensions. Objects have length, width, and depth. You know of the north–south axis, the east-west axis, and the up-down axis. And while for a time video games were constrained to but two of these axes, modern games are often highly immersive experiences in a virtual three-dimensional world.

But for some, three dimensions aren't enough. That's why Marc ten Bosch set out to create a *four*-dimensional puzzle-platform game called *Miegakure*.

In order to understand what we mean by a four-dimensional game, let's first take a look at some analogues of ten Bosch's concept in three dimensions. Just as it's difficult for three-dimensional beings like us to think about a fourth dimension, it would be difficult for two-dimensional beings to think about a third. This is the central conceit behind the nineteenth-century satirical novella *Flatland,* a story about a square in a two-dimensional universe who comes to realize the existence of a third dimension (Abbott 2006). The book has inspired a number of mathematical homages and is a great introduction for anyone who wants to think about higher dimensions.

For a more modern approach, a number of relatively recent games explore similar ideas. (Given the history of video games and their transition from two- to three-dimensional gameplay, this shouldn't be terribly surprising.) One of the more well-known examples is an indie game called *Fez,* originally released in 2012.[22]

In the game, an eight-bit aesthetic combines with more modern gameplay. The player controls a character named Gomez through a two-dimensional world. Early on, however, Gomez discovers the titular fez, which shows him that his world is really just one slice of a three-dimensional reality. With the fez atop his head, Gomez can rotate his perspective and move between different 2D planes.

Other games have explored similar game mechanics. Even video game A-listers like Mario have explored hybrid 2D/3D worlds. For example, in *Super Paper Mario,* released for the Wii in 2007, Mario lives in a 2D world reminiscent of his earliest adventures. Early on, though, he gets the power to rotate into a third dimension. As in *Fez,* this power is used extensively to explore the world, solve puzzles, and find secret items.

In both cases, players take on the role of a two-dimensional character (I'm speaking technically here, not pejoratively). But the central gameplay mechanics—both in *Fez* and in *Super Paper Mario*—let the player explore different 2D slices of a larger, three-dimensional world.

Miegakure is merely an extension of this idea. In this game, the player can explore different 3D slices of a larger, four-dimensional world. This allows the player to do things that we'd never be able to do in our own world. For instance, ten Bosch gives an example of how you can use 4D space to walk through walls. Figure 1.7 provides the necessary visual details.

(a) Wall impeding your progress? No problem!

(b) Rotate through the fourth dimension to a different 3D slice...

(c) Rotate again, and cross the desert...

(d) Rotate again, and you'll see the person on the other side of the wall.

(e) Rotate one final time, and you'll find yourself on the other side of the wall. But really, pictures don't do this process justice. You should go to miegakure.com and check out the video showcasing this sequence of steps.

FIGURE 1.7. Walking through walls using four dimensions. Screenshots from *Miegakure*, reproduced with permission from Marc ten Bosch.

Confusing? This is really not so different than what one might do in *Fez* or *Super Paper Mario* in order to move to the other side of a 2D wall (or chasm, as the case may be). In a promotional video for the game, ten Bosch even showcases a more direct 2D analogue of this scenario

(a) What looks like a 2D world...

(b)...may just be a slice of a 3D one.

(c) Rotate the character's view, and move left to pass into a different 2D slice.

(d) Rotate back and cross the desert.

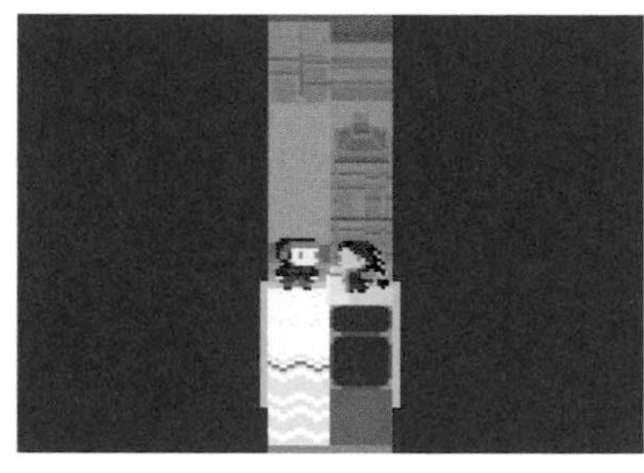

(e) Rotate again, and you'll see the person on the other side of the wall.

(f) Rotate one final time, and you'll be on the other side of the wall.

FIGURE 1.8. Walking through 2D walls using three dimensions. Screenshots from *Miegakure*, reproduced with permission from Marc ten Bosch.

(Figure 1.8). This analogy highlights a mechanic very similar to the central mechanic in *Fez*.

This analogue of 2D planes in 3D space should be easier to follow than 3D spaces in a 4D hyperspace. But that's precisely the point. A game like *Miegakure* offers us something that no other medium really can: the opportunity to build an *experiential* understanding of the fourth dimension. Though mathematicians have been thinking about higher dimensions for hundreds of years, their thought experiments are decidedly less immediate than a game that lets you actually interact with a 4D world.

In other words, *Miegakure* very clearly plays to the strengths of its medium. This fact is not lost on ten Bosch. When I asked him about the educational value of a game like *Miegakure*, here was his response:

> Every game is educational, the only question is *what does it teach you*? *Miegakure*'s difference is that it happens to be based on, and hence teach, the fourth dimension.
>
> One can understand something in multiple ways, such as intuitively (from experience), or symbolically (mathematically). I wouldn't necessarily say that intuitive or symbolic understanding is a "deeper" kind of understanding than the other. Both are useful for different things.
>
> Games are naturally better at teaching non-verbally and giving intuition for a concept through interaction than at teaching the mathematical symbols we use to represent and understand it.

Let's be clear: *Miegakure* is, first and foremost, a game. The fact that it intersects with interesting mathematics is great, but the goal here is to make a great game, not a great educational resource. Even so, when I asked ten Bosch to wax aspirational on his goals for the game, he had this to say:

> At the very least I would like *Miegakure* to move the video-game medium forward. Games have a great potential that has not been reached so far. I think it can be an excellent example of what games can do. I would love if the game can inspire game designers to be more ambitious. It's a game about something fundamental but that we cannot experience without a computer. That also makes it a kind of "educational" game but in a different way than is usually meant. If lots of games using these ideas are created it might change the world a bit. Someone once said to me: "I think some kid is going to play your game and grow up and become a better physicist because of it."

Game design always has educational consequences, because players are always learning something when they play. But it's when the game world intersects with other worlds in meaningful ways that video games really start to show significant potential, especially in contrast to other mediums.

1.7 BEYOND PHYSICS: *MINECRAFT* OR MINE FIELD?

While many games strive for realistic physics, not all do. Moreover, realism isn't necessarily a prerequisite for a game to be used as an educational resource. Nowhere is this clearer than in the case of *Minecraft*.

Originally released in 2011, *Minecraft* throws players into a pixelated world composed of blocks. There's no plot to speak of; instead, players are meant to explore the world around them, *mining* from the blocks to collect things like wood or metal, and using those items to *craft* materials and build structures within the world. The physics is entirely unrealistic—for instance, you chop down trees one block at a time, and removing the trunk of a tree leaves the rest of it hovering in the air—but realistic physics isn't a focal point of the game.

Since its release, the game has become a massive success and was purchased by Microsoft in 2014 for $2.5 billion. Like *Portal* and *LittleBigPlanet*, *Minecraft*'s success hasn't been limited to the consumer market. Educators see a lot of potential in the game too. In late 2011, a team of teachers created a classroom-customized version of *Minecraft* called *MinecraftEdu*, which was rebranded as *Minecraft: Education Edition* in 2016. Similar to *Teach with Portals*, the goal here is to package *Minecraft* with additional resources to make it more suitable for classroom use.

But how exactly is *Minecraft* being used in class? For math class in particular, one example can be found in "The Minecraft Generation," a 2016 article written by Clive Thompson for *The New York Times Magazine*. At one point in the article, Thompson describes redstone, a material in *Minecraft* with conductive properties that simulate electricity:

> Switches and buttons and levers turn the redstone on and off, enabling players to build what computer scientists call "logic

> gates." Place two Minecraft switches next to each other, connect them to redstone and suddenly you have what's known as an "AND" gate: If Switch 1 and Switch 2 are both thrown, energy flows through the redstone wire. You can also rig an "OR" gate, whereby flipping either lever energizes the wire.
>
> These AND and OR gates are, in virtual form, the same as the circuitry you'd find inside a computer chip. They're also like the Boolean logic that programmers employ every day in their code. Together, these simple gates let Minecraft players construct machines of astonishing complexity.[23] (Thompson 2016)

In other words, players can use redstone to learn about the basics of mathematical logic: A AND B is true if and only if both A and B are true, A OR B is true as long as A and B are not both false, and so on. This may sound fairly basic, but with these fundamentals you can build some fairly sophisticated machinery. Modern electronics are based on logic gates, which take binary inputs (e.g., true or false) and return one output.[24]

Of course, in the same way that Legos aren't necessarily an educational resource, neither is *Minecraft*. If it were, there would be no need for an education edition of the game. While using the game as a classroom resource may help get buy-in from students, structure is still required to ensure that what they're learning is aligned with the teacher's instructional goals. Otherwise, educators may find scenarios like the one described in Greg Toppo's book, *The Game Believes in You: How Digital Play Can Make Our Kids Smarter*. While describing one particularly devoted *Minecraft* player, Toppo writes, "At one point, he spent about sixty hours digging the deepest hole possible—he simply dug until the game would let him go no further" (Toppo 2015, p. 107). I won't say that this process yielded *no* learning, but was it the most efficient use of time? Probably not.

1.8 CLOSING REMARKS

Video games have become increasingly complex over the years. But this doesn't mean that they have necessarily become more realistic. Nor is there an expectation that realism is the ultimate goal. Just as every painting need not be photorealistic, there's plenty of room in the marketplace for games with different aesthetics.

Regardless of the level of realism, though, the question remains: Can games be useful educational resources? I hope that by the time you finish this book you'll think the answer is yes, though the process of extracting educational value out of games is not always straightforward. Certainly older games, though they can inspire some fun questions, are also riddled with misrepresentations of how the world works. And while more modern games may have greater potential, the question of how to integrate educational goals with gameplay is still fairly open.

One thing's for sure: grafting educational features onto games not specifically designed for education probably isn't the best way to go. Students should be learning because they want to learn and because they have genuine interest in the topic at hand, not because multiplying large numbers will let them advance further in a level. This sort of artificial incentivizing of material that actually has compelling applications doesn't help teachers or students answer the "When will I ever use this?" question.

It should also be noted that using games in education isn't a panacea simply because people would rather play video games than study. The medium ought to serve the educational goals, and designers interested in education should ask whether the ideas they have for games really provide anything new. Watching a virtual version of a chemical reaction in *LittleBigPlanet* pales in comparison to seeing the same reaction on YouTube.[25] Of course, seeing it in person would be even better, but doing so involves safety risks and requires resources an educator might not have. Nevertheless, video games ought to play to their strengths. Just because a student would rather play a game, doesn't mean she will learn any better if course material is sprinkled inside of a game world. This is especially true of games like *Minecraft,* which have captured a fair amount of popular attention. Taking a boring lesson and putting it in a video game doesn't mean the lesson will become better; usually it just means that the experience of playing the game becomes worse.

Having said that, in some cases a game may be the best way to explore some physical or mathematical phenomenon. Whether exploring near-light speeds or higher dimensions, games provide opportunities to experience and build intuition for ideas in a way that other mediums simply can't. *A Slower Speed of Light* provides a prototype for the potential of the medium. *Miegakure* moves the needle forward even further, by trusting in good design rather than setting explicit educational goals.

I hope that these are issues that designers will continue to grapple with. Will we ever see Mario traveling near the speed of light, or into higher dimensions? Given his more recent sojourns into distant galaxies, I'd like to think that the answer is yes.

1.9 ADDENDUM: DESCRIBING DISTORTION

In this section we'll briefly describe *why* length contraction occurs. The math doesn't go beyond the high school level, but if you'd rather take length contraction at face value, feel free to move on to the next chapter.

To talk about length contraction, it's actually best to start with another strange feature of relativity, known as *time dilation*. Moving near the speed of light doesn't just make distances behave strangely; it makes time behave strangely too.

To understand how time dilation works, imagine that you and I each have cars equipped with special clocks. They work like this: a light beam travels vertically between two mirrors. Each time the light beam completes a round trip from one mirror to the next, our clocks mark off one unit of time. This idea is illustrated in Figure 1.9.

Of course, the distance the light travels depends on the size of our clocks. In general, for an object traveling at constant speed, we can relate its speed, distance traveled, and time traveled by the formula:

$$\text{distance} = \text{speed} \times \text{time, or}$$

$$\text{time} = \text{distance} \div \text{speed.}$$

In the case of our clocks, suppose that the two mirrors are separated by a distance d. Let's let c stand for the speed of light, and t be the time it takes for the light to travel from the bottom of the clock, up to the top of the clock, and back down again. Since this path covers a distance of $2d$, the distance formula above tells us that $t = 2d/c$.

With one of these clocks in each of our cars, we're able to precisely measure time (in theory). Now, imagine that since we're such good

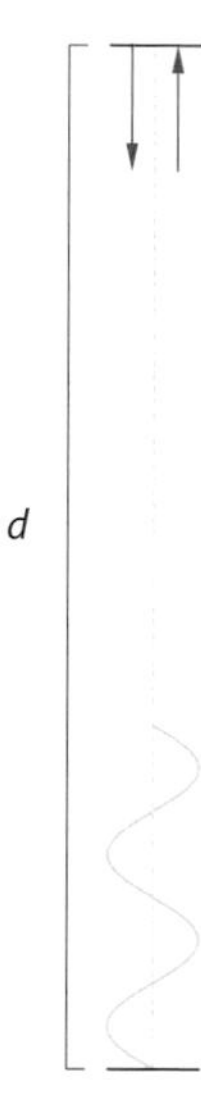

FIGURE 1.9. Image of our light clocks. The yellow beam of light moves to the top, hits the mirror, bounces off, hits the bottom mirror, and repeats.

friends, we'd like our clocks to be synchronized, so that the clocks in our cars will read the same time and one of us won't inadvertently keep the other one waiting when we make plans to meet up.

After some deliberation, we decide that the best way to synchronize our clocks is to do it one morning when you are on your way to work, since my house is on your route. I wait with my car parked on the street, and the moment you zip by me, we both set our clocks to the same time. Nothing could be simpler, right?

Well, maybe not. If you're driving at a constant speed v while I'm at rest, then from my perspective the light in your clock will have to travel *farther* as it bounces between the two mirrors. This is because your beam not only has to move up and down, but it also has to maintain the car's horizontal speed.

Let's say you're driving to the right. From my perspective, the path of the light beam inside your clock must look something like Figure 1.10. And since it appears that your light beam has to travel *farther*, your clock will appear to be running *slower* than mine.

We can do a bit better than this; in fact, I can tell you exactly how much slower I think your clock is. Let's let t^* denote the time it takes for

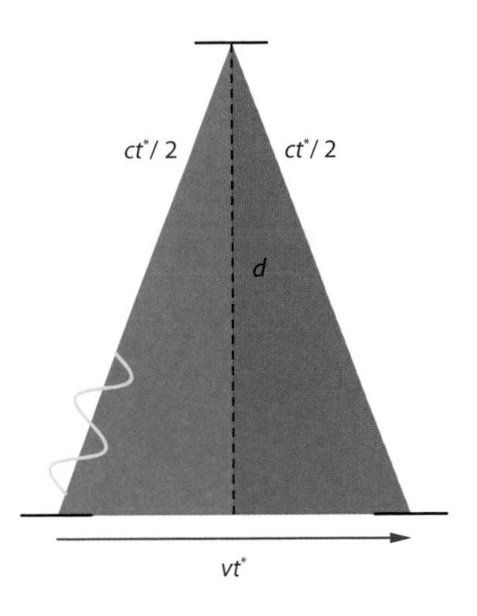

FIGURE 1.10. Image of your moving light clock from my perspective.

the light beam to make one round trip in your clock. Since the speed of light is c, the distance the light travels in time t^* is ct^*. In other words, the light travels a distance of $ct^*/2$ to get from one mirror to another, and another $ct^*/2$ to return back to where it started.

Moreover, since your car is moving at speed v, the bottom mirror has traveled a distance of vt^* by the time the light beam returns. Taking all of this into account, your traveling clock forms two congruent right triangles (the red and blue ones in Figure 1.10), both of which have sides of length $vt^*/2$, d, and a hypotenuse of length $ct^*/2$. By the Pythagorean theorem, this means that

$$d^2 + \left(\frac{vt^*}{2}\right)^2 = \left(\frac{ct^*}{2}\right)^2.$$

This is equivalent to the equation

$$(t^*)^2\left(c^2 - v^2\right) = 4d^2 = c^2t^2,$$

since we saw above that $d = ct/2$. By taking positive square roots and isolating t^*, we reach the equality

$$t^* = \gamma t, \quad \text{where } \gamma = \frac{1}{\sqrt{1 - v^2/c^2}}.$$

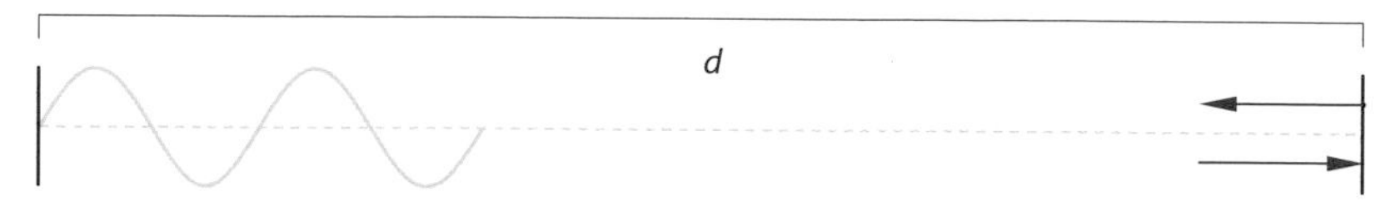

FIGURE 1.11. Light clocks in our new configuration.

In other words, my clock is γ times faster than yours: every second on your clock corresponds to γ seconds on mine, and γ is always greater than 1. This is as precise as we can get when it comes to my claim that your clock is running slow.

It takes less work to derive the formula length contraction once we've got the formula for time dilation. To see why length contraction occurs, imagine the same scenario as before, but with our clocks positioned on their sides. Instead of our mirrors being on the top and bottom, they're now on the left and right.

Of course, how we measure time doesn't depend on the orientation of our clocks, and since your clocks run slow (from my perspective) in the first configuration, they'll continue to run slow in the second. The difference now, of course, is that there are no triangles to consider, because the light travels parallel to the motion of the car (Figure 1.11).

Since the speed of light is the same in both of our cars, but the times we measure are different, the distance light travels must be different in each vehicle. More specifically, if the distance between the mirrors in your car is d^*, then in order for one tick on your clock to correspond to γ ticks on my clock, d^* must be shorter than d by a factor of γ. In other words, $d^* = d/\gamma$.

When γ is close to 1, which is exactly what happens when the velocity of the moving object is small compared to the speed of light, these effects are negligible. But as γ gets larger—that is, as your car's velocity gets closer and closer to the speed of light—length contraction has an increasingly dramatic effect.

In keeping with the spirit of *A Slower Speed of Light*, imagine light traveled at 100 mph. Then, if you drove by me in your blue car at 70 mph while I was at rest in my red car, I'd see the scenario depicted in Figure 1.12. However, from your perspective things would look like the image in Figure 1.13.

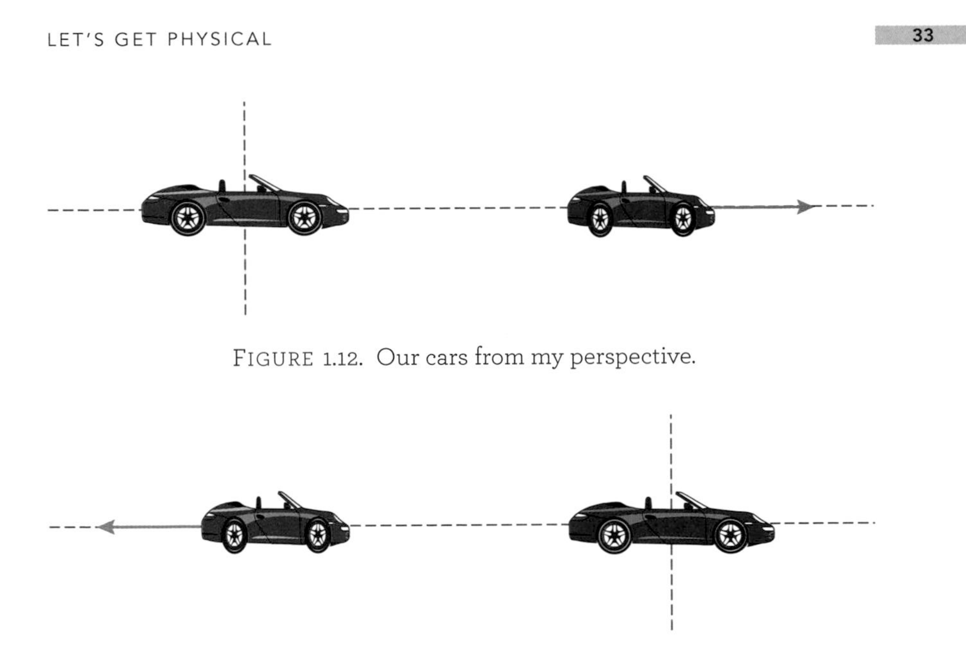

FIGURE 1.12. Our cars from my perspective.

FIGURE 1.13. Our cars from your perspective.

When you move and I stay at rest, from my perspective your clocks run slow and your lengths are shortened. Of course, from *your* perspective it's the opposite: my clocks run slow and my lengths are shortened, since in your frame of reference you're at rest and I'm moving away from you. When it comes to time dilation and length contraction, it's literally all relative.

2
Repeat Offenders

2.1 LET'S PLAY THE FEUD!

As a child, I probably watched more than my fair share of game shows on TV. In the evening we watched *Jeopardy* as a family, and on days when I stayed home sick, *The Price Is Right* was must-see TV. In high school, friends and I would even occasionally duck into an empty AV-equipped classroom if *The Price Is Right* was on during one of our free periods. As you can imagine, we were all quite popular with the ladies.

Family Feud was another stay-at-home staple. The show's popularity has ebbed and flowed throughout the years, but since Steve Harvey became the host in 2010, *Family Feud* has become more popular than ever. In fact, 2015 marked the first time that *Family Feud* became the most popular game show on television (as measured by Nielsen ratings), surpassing both *Jeopardy* and *Wheel of Fortune*.[1]

If you've never seen the show, here's how it works: two five-person teams compete for points by trying to guess the most popular responses to survey questions from a sample of 100 people. For example, the host might ask, "What's the first thing you do after waking up from a nap?" Contestants will then try to guess the most popular answers, which in this case could include going to the bathroom, checking the time, or yawning.

Teams earn points for every response they guess correctly. Once a team wins (either by getting the most points, or by reaching a certain point threshold), that team moves on to the "fast money" round to compete for a cash prize.

For me, it was always fun to guess the most popular responses before the complete list was revealed. Watching contestants make terrible guesses was another source of joy (just search on YouTube for "Family

Feud fail" to see what I mean). So you can imagine my adolescent excitement when I discovered a video game version of this show during a routine visit to my local video store. A game that captures the experience of playing *Family Feud*—how could it be anything less than amazing?

Well, I'll tell you how. To begin with, the game was for the Super Nintendo, a system with no keyboard and no autocomplete. This meant that in order to type in a word or phrase, you had to select letters from an on-screen keyboard by moving from letter to letter, one character at a time. Giving an answer was a much bigger pain in the virtual edition of the game.

A more significant problem concerned the game's intelligence, or lack thereof. It always seemed like for every answer the game interpreted correctly, there were two that it completely whiffed. Here's an example of what I mean. In the game, "What's the first thing you do after waking up from a nap?" is an actual question. Here are the answers (along with the number of people who responded with that answer):

1. Go to bathroom (21 responses)
2. Check the time (18 responses)
3. Stretch (15 responses)
4. Yawn (9 responses)
5. Get a drink (8 responses)
6. Wash face (8 responses)
7. Eat (4 responses)
8. Brush hair (3 responses)

If you happen to stumble upon this question while playing the game, Box 2.1 highlights some things you may notice.

In the grand scheme of things, problems with typing and the game's intelligence are really just limitations of the technology. After all, the game was released in 1993, and it's a little unfair to judge it by today's standards. Modern versions of the game do exist: new *Family Feud* games were released in 2009, 2010, and 2011. There are also a couple of more recent mobile versions of the game.

These later games can still be a bit finicky when it comes to giving you credit for answers, but things have improved considerably over 20 years. Typing guesses is relatively painless now, too: mobile versions of the game use a smart keyboard, which attempts to guess what you're typing as you type it. (Of course, this painlessness comes at a cost, since a keyboard using autocomplete may inadvertently reveal good guesses that the player otherwise might not have considered.)

INTELLIGENT BEHAVIOR
• Typing "use toilet" will give you credit for "Go to bathroom," even though the two phrases have no words in common.
NOT-SO-INTELLIGENT BEHAVIOR
• Typing "comb hair" gives you no credit for "brush hair." • Typing "brush teeth" gives you credit, not for "brush hair," which might make some sense, but for "eat."

Box 2.1. Quirky behavior in *Family Feud.*

There is a deeper problem, though. It's one that any player will notice, and it becomes worse the more frequently one plays. At its root it is a fundamental mathematical truth, and so unlike the problems mentioned above, technology has a limited ability to fix it. And unfortunately, the issue isn't limited to games based on *Family Feud*; it affects a large number of games, and may have even cost one popular mobile gaming company hundreds of millions of dollars.

In what follows, we'll describe this problem, discuss the effect it has had on a handful of games, and explore the difficulties inherent in trying to come up with a solution.

2.2 GAME SHOWS AND BIRTHDAYS

To understand the nature of this problem, let's return to the Super Nintendo version of *Family Feud*. The game has around 2,000 questions to draw from, though they are placed into different categories. There are 498 questions in the "single-point" round, 481 questions in the "double-point" round, 501 questions in the "triple-point" round, and 496 for the final speed round.[2] To simplify things, let's just assume that questions in each round are drawn from a pool of 500.

Because the number of questions is fixed, as you play the game, you are bound to encounter questions you've already seen. The more you play, the more frequently these duplicates will come up. And because the game reveals all the answers to each question at the end of the round—regardless of how many you guessed correctly—when these questions repeat themselves, you may remember some of the answers. Over time, then, the game becomes increasingly familiar, and much of the fun disappears.

So what? You may be thinking. After all, 500 questions per round is a lot. It will probably take a while for the repeats to start showing up, and once they do, it's not like you'll be able to instantly recall every correct response.

All of that may be true. On the other hand, once the repeats start showing up, they will appear with greater and greater frequency. More important, the repeats are likely to show up sooner than you think.

How much sooner? Well, let's assume that each question is equally likely to be chosen. What's the probability that the first question you draw will be one you haven't seen before? That's easy; it's 500/500 = 100%, since you haven't seen any questions.

The probability that the second question will be one you haven't seen before is 499/500 = 99.8%, since the only way for you to get a duplicate is for the second question to be the same as the first.[3] If your first two questions are different, the probability that your third will be different is then 498/500, or 99.6%, since there are still 498 questions you haven't yet seen. Since your first three questions will have no duplicates if your first question isn't a duplicate, *and* your second question isn't a duplicate, *and* your third isn't a duplicate, the probability that you won't see a duplicate among the first three questions is the *product* of the probabilities above:

$$\frac{500}{500} \times \frac{499}{500} \times \frac{498}{500} \approx 99.4\%$$

This probability is certainly high, but don't let that fool you. Suppose you want to know the probability that you won't see a duplicate among your first k questions, for some value of k. This happens precisely when your second question is different from the first, your third question is

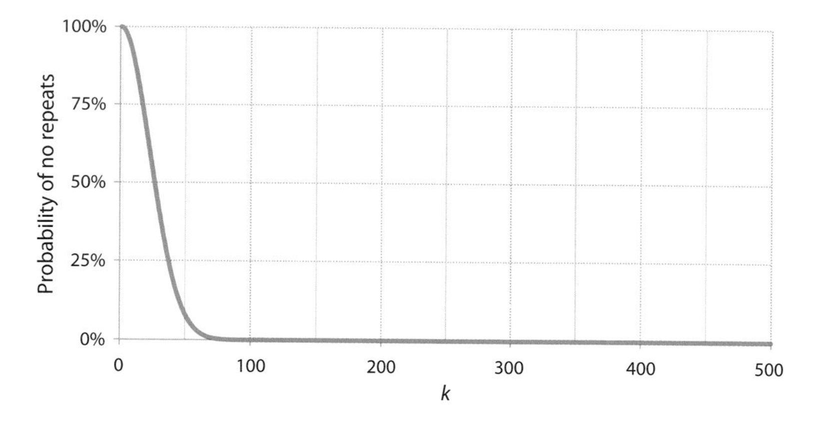

FIGURE 2.1. As the number of rounds (k) increases, the probability of not seeing any repeats decreases.

different from the first two, your fourth question is different from the first three, and so on. By the same reasoning as before, this probability equals[4]

$$\frac{500}{500} \times \frac{499}{500} \times \ldots \times \frac{500-(k-1)}{500}.$$

Let's explore how this quantity changes as k increases. Figure 2.1 shows a graph of this expression for every possible value of k from 1 to 500. In this case, it takes only 27 questions for the probability of not getting a duplicate to drop below 50%[5]. That's barely more than 5% of the total question pool! Similarly, the probability of not having a duplicate among the first 38 questions is less than 25%. Among the first 48 questions, the probability drops to less than 10%, and among the first 67 questions it drops to less than 1%. Simply put, you're very likely to encounter a duplicate relatively early.

If you've taken a course in probability, this phenomenon may sound familiar to you. It's more commonly known as the "birthday problem" and is typically formulated like this: what's the smallest number of people you need to gather together to have at least a 50% probability that two of them will share the same birthday? Replace "people you need to gather together" by "questions you need to see in the *Family Feud* game" and "will share the same birthday" with "be the same question," and you have the problem we've been talking about.

Since there are roughly 365 days in a year, versus roughly 500 questions in each round of *Family Feud,* it should come as no surprise that the number of people needed to have at least a 50% probability of a duplicate birthday is smaller than the number of questions needed to have at least a 50% probability of a duplicate. But it's not that much smaller: only 23 people are required for the birthday problem, versus 27 for the *Family Feud* version.

In both cases, the fact that duplicates appear with such high probability so early is often met with some skepticism. After all, many of us know more than 23 people, while fewer of us know someone who shares our birthday. For instance, if your birthday is on January 1, the probability that someone you meet will share your birthday is around 1/365; put another way, you should expect to have to meet around 365 people before you meet someone who shares your birthday.

However, the birthday problem doesn't ask how likely you are to find someone with the same birthday as *you*; it merely asks how likely you are to meet two people with the same birthday. This is a weaker condition because it doesn't impose the date that must be shared. The only important thing is that *some* date is shared by two people. When people think the solution to the birthday problem should be a much larger number, it's often because of this misconception.

2.3 BEYOND THE FIRST DUPLICATE

When the game requires you to draw from a finite list, we've seen that you're likely to find your first duplicate relatively early on. But this isn't the only interesting mathematical feature of these games. Let's explore a couple more: how many duplicates you're likely to see as you play the game more, and how likely your *next* draw is to be new.

Suppose we play through k single-point rounds of *Family Feud,* for some value k. How do we expect the number of repeats to grow as we go through more rounds (i.e., as k increases)? For later discussion, it will help to have a more general framework, so let's also suppose there are N different questions to pull from, instead of 500. The expected number of repeats among our k questions will depend on both k and N. Let's call this number $R(k, N)$.

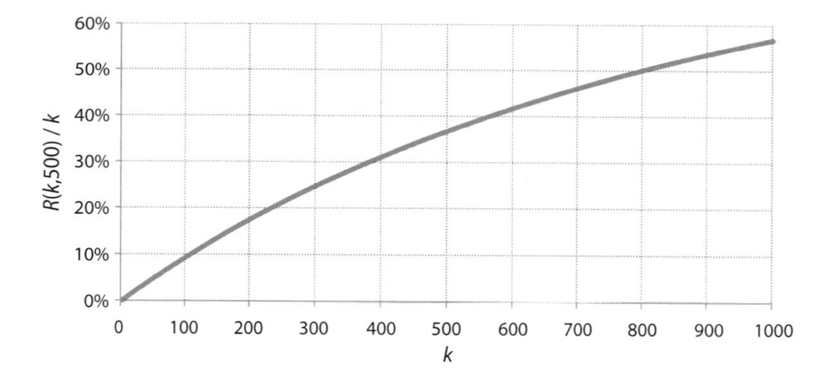

FIGURE 2.2. As the number of rounds (k) increases, expected number of repeats increases too.

With a little work, we can show that[6]

$$R(k, N) = k - N\left(1 - \left(1 - \frac{1}{N}\right)^k\right).$$

Let's look at a specific example. Suppose once more that there are 500 different questions to pull from, and that you've played 50 rounds. At this point in the game, you can expect to have seen $R(50, 500) \approx 2.4$ repeats. In other words, among the first 50 rounds you should expect slightly less than one repeat per 20 questions, for a failure rate of around 5%.

And of course, this percentage only grows as k increases, since the more questions you see, the more duplicates you see, and each new question decreases the number of new questions remaining by one. Unsurprisingly, the expected number of duplicates as a percentage of the total number of questions seen (that is, the ratio of $R(k, 500)$ to k) grows as k increases (Figure 2.2).

As we've already discussed, after 50 questions you can expect around 2.4 duplicates, or roughly 5% of the total. After 73 questions, you can expect around 5 duplicates, or around 6.8% of the total. And after 104 questions, you can expect around 10 duplicates, or slightly less than 10% of the total. While these percentages don't grow terribly quickly as k increases, they do grow. This is bad news for everyone involved, since players are less likely to return to the game if they feel like it keeps serving up old content.

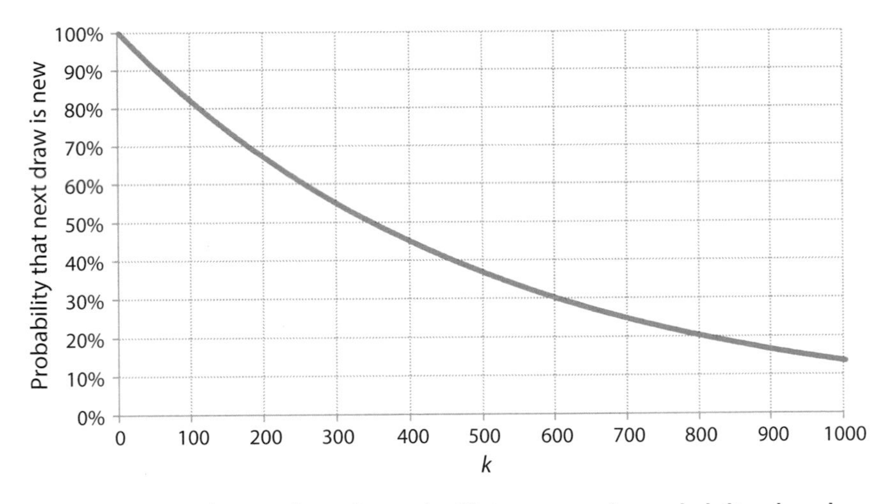

FIGURE 2.3. As the number of rounds (k) increases, the probability that the next round is new decreases exponentially.

The bad news is only compounded if we look at how likely it is for your *next* draw to be new. Again, let's suppose there are N questions to pull from. After k draws, the probability that your next card is new is given by the expression[7]

$$(1 - 1/N)^k .$$

In other words, the probability that your next card will be new decreases *exponentially* with the number of rounds you've already played, regardless of the size of the question pool! No matter what, it seems like you're bound to encounter a fair number of duplicates as you play (Figure 2.3).

2.4 THE *DRAW SOMETHING* DEBACLE

But still, so what? Who cares about a 1993 Super Nintendo game based on a silly daytime TV game show? Well, as with any good bit of mathematics, the analysis performed here extends well beyond the present example. For starters, one could easily analyze other video game versions of *Family Feud*. More recent adaptations of this show have thousands more questions but still must face up to the mathematical consequences of their finite question lists.

Other games based on popular shows suffer as well, like *Jeopardy* and *Wheel of Fortune*. Older games had many of the same problems as the early *Family Feud* games, but even more modern versions must combat the diminishing return on enjoyment that comes from exhausting the supply of new questions (or new answers, whatever the case may be).

The issue of repetition doesn't pop up just for games based around TV game shows; it becomes a headache for big-budget sports games as well. For instance, each major sport has a corresponding video game franchise: for football, there's the hugely successful *Madden* franchise; for baseball, there's Sony's *MLB: The Show*; and for basketball, the *NBA 2K* franchise reigns supreme. Each of these games operates on an annual release schedule, with the latest games offering the best graphics, most-up-to-date rosters, and improved audio.

Audio is where things get tricky. In modern sports games, announcers record hundreds of thousands of lines of dialogue to make for a more immersive and authentic experience. This is in stark contrast to sports games of the late '80s and early '90s, when people were practically blown away by *any* amount of recorded dialogue in a game. According to NPR's *All Things Considered*, a basketball game released for the Sega Genesis during that time may have had 75 lines of recorded dialogue, compared to more than 250,000 lines in a sports game released today.[8]

But even though the amount of dialogue seems quite large, it's still finite. Unlike real broadcasting, there's no room for improvisation. And nearly all of the dialogue is context-dependent: the things an announcer would say at the very end of a close game are different from the things an announcer would say at the end of a blowout. This means that even with a quarter of a million lines of recorded audio, the number of options for what an announcer can say in any given scenario may be quite limited. That's why, even in the most recent iterations of these popular franchises, you may notice the announcers repeating themselves after only a single play session.

Fortunately, repetitive announcers in sports games don't have a significant effect on the gameplay itself. Even if you've heard everything that the announcers have to say, you can still enjoy playing a virtual game of football. But for other games, this may not be the case.

Perhaps nowhere is the negative impact of duplicate questions more apparent than in the case of *Draw Something*. A spiritual successor to the classic drawing game *Pictionary*, *Draw Something* was released on mobile devices in February 2012 and, less than two months later, had been downloaded more than 35 million times and had 15 million daily active users. In the game, players were presented with a handful of words. They would select one, draw a picture representing the word, and send it to a friend. The friend tried to guess what had been drawn, and then drew an image of his or her own.

This visual back-and-forth proved to be highly addictive. That addictiveness translated into a huge payday: by the end of March, the company behind the game, OMGPOP, was purchased by social game behemoth Zynga for nearly $200 million.[9] By any measure you can think of, the game was an instant success.

Unfortunately, their success proved to be short-lived. After the number of daily active users peaked at the beginning of April 2012, it plunged more than 40% in the span of just thirty days, from around 15 million to 9 million. While social games have a tendency to come and go (*Farmville*, anyone?), *Draw Something*'s fall from grace was particularly sudden.

It was not entirely surprising, however. In fact, Justin Davis wrote an article for gaming website IGN almost a week after the Zynga purchase was announced, titled "Why *Draw Something* Blew Up, but Might Fade Fast." His biggest concern about the game's long-term prospects? The number of repeated words. In the article, Davis wrote:

> *Draw Something*'s word pool is too small. Way too small. Anyone that consistently puts in time for even a few days will start seeing repeats. This breaks the game, plain and simple. And if you play as seriously as some Editors around the IGN office, by now *most* words are repeats ... To keep the app's most ardent fans from moving on, *Draw Something*'s word pool needs to be 5X its current size (Davis 2012).

How bad was it for the average player? Let's do some quick estimation. A March 1, 2012, article written about the game claimed that at that point there were 6 million users, and 1,000 pictures were being uploaded every second (Kessler 2012). At that rate, it means users

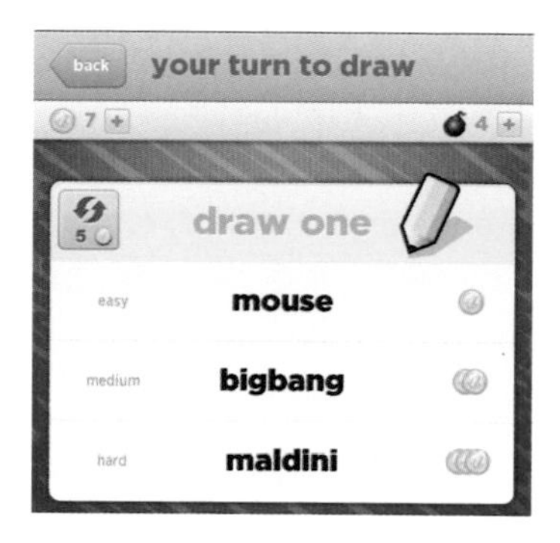

FIGURE 2.4. Example of the choices presented before a player draws. These three options will always be grouped together. Screenshot from *Draw Something*, reproduced with permission from Zynga.

uploaded a whopping 86.4 million pictures per day, for an average of 14.4 daily picture uploads per user. This may sound like a lot, but if you had five friends who played the game daily during *Draw Something*'s heyday, it amounts to fewer than three pictures sent to each friend per day. If that still sounds high, we can round things down to 12 daily picture uploads per user. Extrapolating across thirty days, this amounts to the creation of around 360 drawings in a month.

This is a nice estimate, but on its own it won't tell us how many duplicates an average user could have expected to see over a thirty-day period. For that, we need to know the size of *Draw Something*'s word pool. During the game's peak, estimates for the number of words hovered around 2,000, though when the game was initially launched the word pool may have been smaller. Let's assume that in March 2012, relatively early in *Draw Something*'s life, the word pool was closer to 1,800.

When a player begins her turn, however, she is not simply given a word; rather, she is given a choice among three words of varying difficulty. But the same three words are always grouped together; for instance, if "bigbang" is your medium word, "mouse" will always be your easy word, and vice versa (Figure 2.4). Because of this, even though there may have been 1,800 words in *Draw Something* in the spring of 2012, there were only 1,800 ÷ 3 = 600 distinct *triplets* of words. This greatly increases the rate at which players would have seen duplicates. Though the average player drew 360 pictures over a thirty-day period, she actually chose from 360 *triplets* of words.

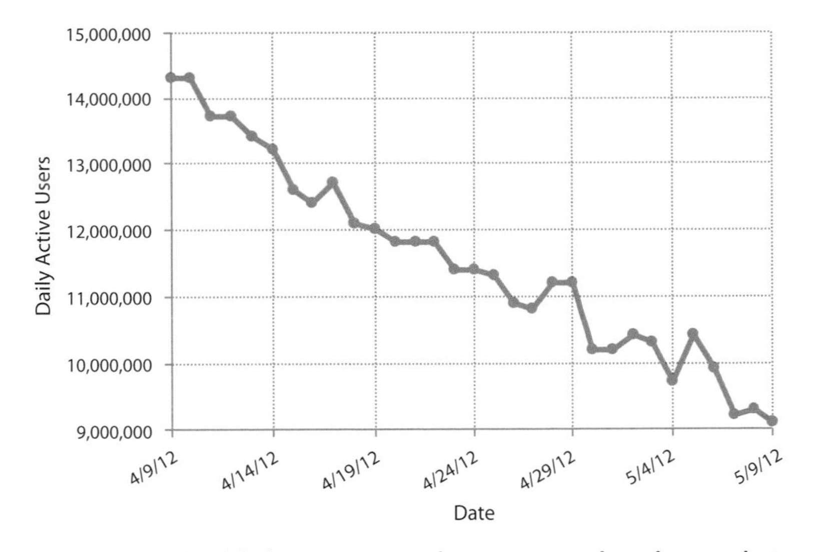

FIGURE 2.5. Graph of daily user activity for *Draw Something* from early April to early May 2012. Source: appdata.com, via Narcisse (2012).

Moreover, players don't just see words when they are drawing. They learn words when they are guessing their friends' words, too! Since all daily active users sent an average of 360 pictures, they also received an average of 360 pictures, coming from $360 \div 3 = 120$ different word triplets. So in addition to the 360 word triplets a typical player saw when drawing, he also saw an average of 120 additional triplets from receiving the pictures of his friends.[10]

This means that out of 600 triplets of words, daily active users over a thirty-day period had likely seen $360 + 120 = 480$ triplets, including duplicates. To find the expected proportion of duplicates, we can use our formula for $R(k, N)$ with $N = 600$. After seeing 480 triplets, we can therefore expect $R(480, 600) \approx 149.4$ duplicates. In other words, more than 30% of the 480 triplets would have been repeats! When you're presented with options you've already seen almost one in three times you play the game, that's certainly going to have an effect on replay value.

Unfortunately for Zynga, the situation continued only to deteriorate (Figure 2.5). A year later, Zynga released *Draw Something 2*, but players had already moved on to the next big thing, and lightning failed to strike twice. In August of 2013, reports surfaced that Zynga was closing down

OMGPOP despite the hundreds of millions of dollars the company had spent just a year and a half earlier.[11]

As for the original *Draw Something*, it's seen a fairly steady decline since its initial release. In fact, *Draw Something* was apparently such a poor draw (pun intended) that in 2014, Zynga rebranded *Draw Something 2* as *Art with Friends*. According to recent estimates, both this game and the original *Draw Something* now have upwards of 7,400 words in their libraries. While an impressive increase, this is still a far cry from the fivefold improvement recommended by Justin Davis in 2012.

Of course, it's impossible for us to infer a causal relationship between the size of the word pool and the number of active *Draw Something* users. There are undoubtedly other factors that contributed to the game's demise. Having said that, maybe if Zynga had done a bit more mathematical due diligence, they could have saved themselves a few million dollars.

2.5 DELAYED REPETITION: INCREASING N

When it comes to gameplay, repetition can be a killer. So, can anything be done to overcome this mathematical obstacle?

Following Justin Davis's advice seems like a good start. If *Draw Something* had had five times as many words in 2012, the number of expected duplicates would have grown much more slowly. Figure 2.6 shows graphs of $R(k, 600)/k$ and $R(k, 3{,}000)/k$. As you can see, the expected number of duplicates grows at a slower rate when you're drawing from a larger pool of items.

For instance, when we increase the number of word triplets by a factor of five, from 600 to 3,000, expected percentage of repeats after a month of use ($k = 480$) drops from around 31.1% to around 7.58%.[12] Similarly, the probability that the next triplet is new decays more slowly (though still exponentially) with a larger word pool.

On the other hand, if you're the type of player to throw up your arms in frustration after the first duplicate, you should cut the developers some slack. To see why, let's return to the birthday problem for a moment and see how our answer changes as the size of the item pool increases.

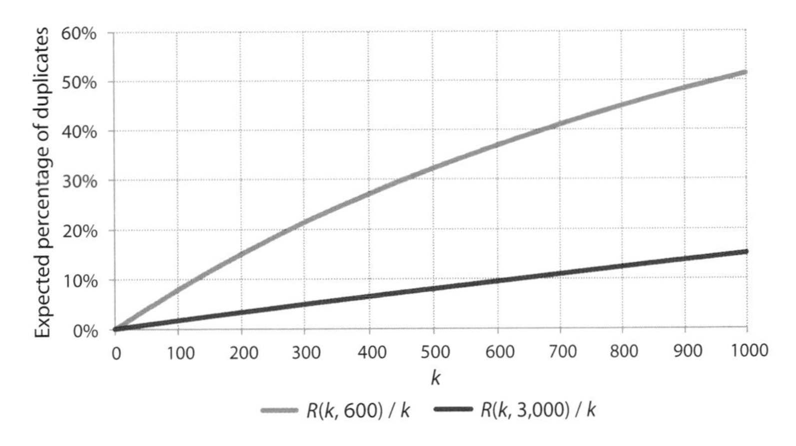

FIGURE 2.6. Effect of increasing the word pool on expected percentage of duplicates. Source: appdata.com.

If we're drawing with replacement from a pool of N items, how many do we have to pull until we have at least a 50% probability of seeing a duplicate? By replacing 500 with N in our argument from the beginning of the chapter, we see that the probability of not finding a duplicate among our first k draws is equal to

$$\left(1-\frac{1}{N}\right)\times\left(1-\frac{2}{N}\right)\times\ldots\times\left(1-\frac{k-1}{N}\right).$$

If we want to have at least a 50% probability of seeing a duplicate, this is the same as saying the above expression should be at most 50%. In other words, given the fact that

$$\left(1-\frac{1}{N}\right)\times\left(1-\frac{2}{N}\right)\times\ldots\times\left(1-\frac{k-1}{N}\right)<\frac{1}{2},$$

we want to understand how N and k are related.

If we take our expression at face value, this seems like an impossible task. The expression is fairly complicated; how could we possibly tease out a simple relationship between N and k?

While a simple *exact* relationship may be hard to come by, a simple *approximate* relationship is within our reach. As we'll show in the addendum to this chapter, a good approximate formula for k is the

Table 2.1. Minimal k for values of N

N	$\frac{1}{2}+\frac{\sqrt{8N\ln 2+1}}{2}$	*Minimal* k
365	23.00	23
500	26.83	27
1,000	37.74	38
2,000	53.16	54
5,000	83.76	84
10,000	118.24	119

smallest whole number satisfying

$$k \geq \frac{1}{2}+\frac{\sqrt{8N\ln 2+1}}{2},$$

where $\ln 2 \approx 0.693\ldots$ is the natural logarithm of 2. This gives us the right answer for the birthday problem and the $N = 500$ case we considered in *Family Feud* (Table 2.1).

The above inequality also tells us that k grows roughly in proportion to the square root of N. This is not terribly fast. Indeed, if you want to double the number of rounds someone will need to play in order to keep the probability of seeing a duplicate below 50%, you need to roughly *quadruple* the size of the pool from which you're drawing! The moral here: for these types of games, despite the best efforts of the developers, you'll probably encounter a duplicate sooner rather than later.

2.6 DELAYED REPETITION: WEIGHT LIFTING

If we wanted to delay the appearance of repeats, increasing the size of the word pool isn't our only option. There are some other things we could do. One approach would be to draw items from the pool *without replacement*—that is, once a question is taken from the pot, it isn't put back in until the pot is empty, at which point *all* the questions are put back in play. In other words, a *Family Feud* game developer could decide that once a player sees a question, she won't see that question again until she's seen every other question first.

This delays the appearance of repeats for as long as possible. Unfortunately, once a player sees a single repeat, it means that everything

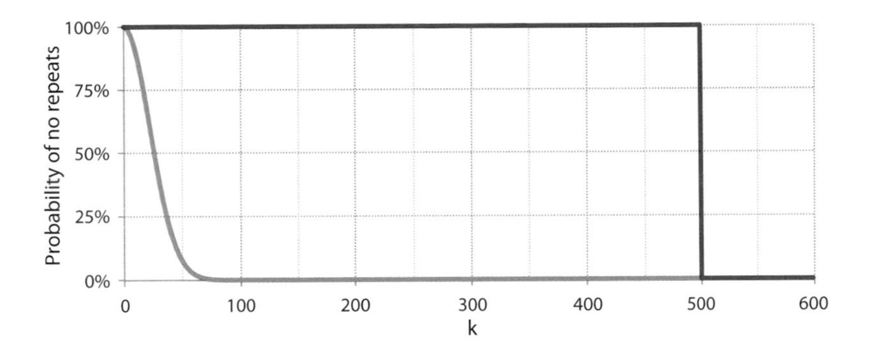

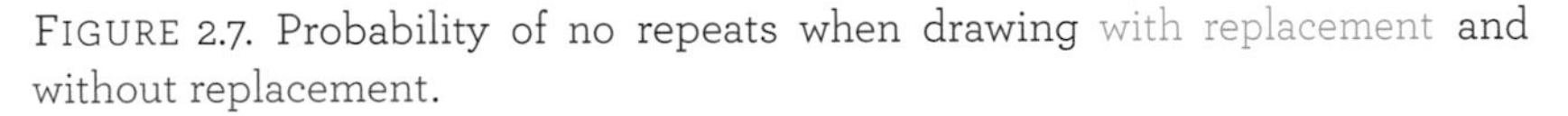

FIGURE 2.7. Probability of no repeats when drawing with replacement and without replacement.

afterward will be a repeat as well. For a more visual spin on this story, Figure 2.7 shows what the number of repeats after k draws would look like in a pool of 500 questions, drawing both with replacement (as we've considered throughout this chapter) and without.

From this picture, a question emerges: is there a middle road? Can we find a curve that sits somewhere between these two, so that the first repeat is delayed with a higher probability, but without going to the extreme of drawing without replacement?

The answer is yes, though such an approach isn't without consequences. One way to compromise between these two extremes is to manipulate the probabilities so that questions which have already appeared are less likely to come up again, and questions which haven't yet appeared are more likely to appear. Indeed, drawing without replacement is just an extreme example of this, since questions that have come up already have a 0% probability of coming up in the future (until all the questions have been drawn).

To see how this might work, let's look at a concrete example. Consider a six-sided die: on your first roll, the probability that the die will land on any one of its six sides equals 1/6. Suppose you roll the die and it lands on 1. Ordinarily, if you rolled the die again, the probability of rolling a 1 would again be 1/6. But imagine that you could alter the die after that first roll, so that the probability of rolling a 1 was half, or even a third of the probability of rolling one of the other values.

If you want the probability of rolling a 1 to be half as likely, it's tempting to say that it should equal 1/12. But this isn't quite right, because the sum of all the probabilities must equal 1, and

$$\frac{1}{12}+\frac{1}{6}+\frac{1}{6}+\frac{1}{6}+\frac{1}{6}+\frac{1}{6}=\frac{11}{12}<1.$$

Instead, if you want the probability of rolling a 1 to be half the probability of rolling one of the other values, then we can find the correct probabilities by letting the probability of rolling one of the other values equal q. In this case, we need

$$\frac{q}{2}+q+q+q+q+q=\frac{11q}{2}=1,$$

so $q = 2/11$. In other words, if you wanted the probability of rolling a 1 to be half as likely as rolling one of the other values, you should set the probability of rolling a 1 equal to 1/11, and set the other probabilities equal to 2/11.

In general, if one value has appeared k times more than another, you could adjust the probabilities by making the more frequent value $(1/2)^k$ times less likely to appear on the next roll, or $(1/3)^k$ times less likely, or p^k times less likely for any $p \leq 1$. We refer to p as the *weight*, since it weighs the probabilities in favor of questions that haven't appeared yet. Table 2.2 shows how this might play out for a few subsequent rolls of the dice in our current scenario, with $p = 1/2$.

While it's not easy to adjust dice-rolling probabilities on the fly, it's certainly reasonable for machines to adjust the probabilities that arise in a video game. If we take this idea of weighing questions, so that ones that have already appeared are less likely to reappear in the future, what sort of picture do we get?[13]

Unfortunately, trying to calculate the relevant probabilities by hand with this approach becomes messy, because the probabilities each round are no longer independent of one another; the probability of any question being picked in the first round is $1/N$, but the probabilities then change as you move from one round to the next depending on what's already appeared. Instead, it's faster to let a computer do the heavy lifting for us, so that's exactly what we'll do.

First, Figure 2.8 shows how the probability of seeing no repeats after the kth question changes when you include a weight as described above. The green line represents a weight of 1/2, while the yellow represents a

Table 2.2. Example of Weighting Dice Probablities in Favor of Values That Have Appeared Less Frequently

Round	*Roll a ...*	*p = 1/2*	*1*	*2*	*3*	*4*	*5*	*6*
0	—	Counts	0	0	0	0	0	0
		Prob. for next round	1/6	1/6	1/6	1/6	1/6	1/6
1	1	Counts	1	0	0	0	0	0
		Prob. for next round	1/11	2/11	2/11	2/11	2/11	2/11
2	1	Counts	2	0	0	0	0	0
		Prob. for next round	1/21	4/21	4/21	4/21	4/21	4/21
3	5	Counts	2	0	0	0	1	0
		Prob. for next round	1/19	4/19	4/19	4/19	2/19	4/19
4	3	Counts	2	0	1	0	1	0
		Prob. for next round	1/17	4/17	2/17	4/17	2/17	4/17
5	5	Counts	2	0	1	0	2	0
		Prob. for next round	1/16	1/4	1/8	1/4	1/16	1/4

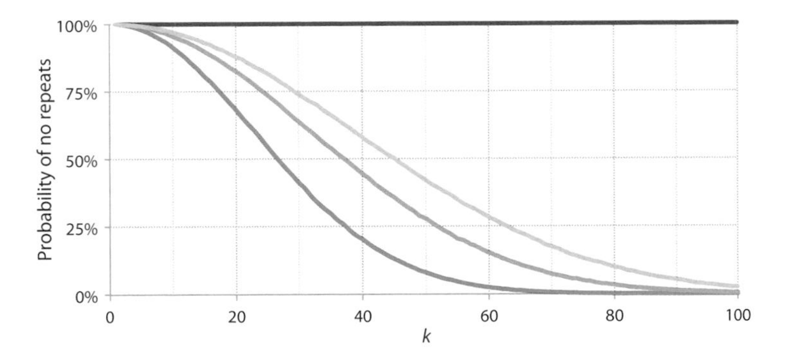

FIGURE 2.8. For each value of k, 100,000 k-round trials were simulated, resulting in the data plotted above. The different graphs correspond to drawing with replacement, with a weight of 1/2, with a weight of 1/3, and without replacement.

weight of 1/3. $N = 500$ for these simulations, just like in the *Family Feud* scenarios from before.

As you can see, these weights work exactly as advertised: they increase the likelihood that players will play longer before seeing their

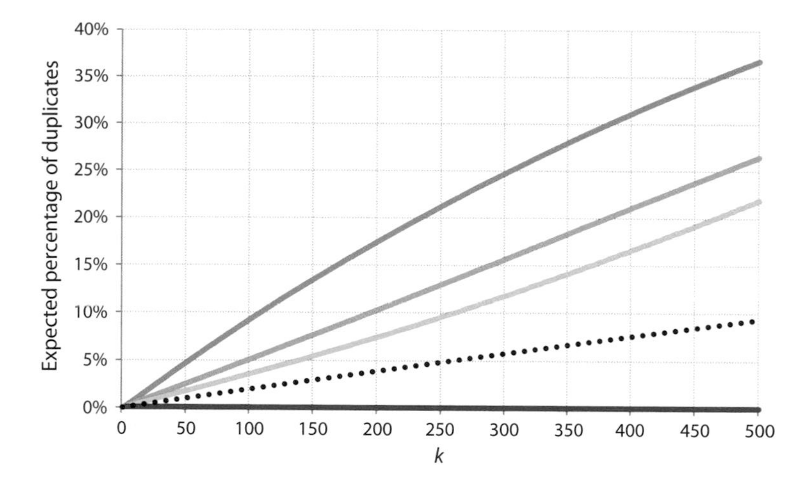

FIGURE 2.9. The expected percentage of duplicates when drawing from a pool of 500 questions with replacement, with a weight of 1/2, with a weight of 1/3, and without replacement. For comparison, the dotted curve represents the expected number of repeats when drawing from a pool of 2,500 questions with replacement.

first repeat, but unlike drawing without replacement, still contain an element of uncertainty about whether any particular round will feature a repeat question.

Similarly, Figure 2.9 shows the expected number of total repeats after k rounds for varying weights p. If increasing the number of questions requires too much of an investment, decreasing the weight can have a similar effect on the expected percentage of duplicates.[14]

However, this doesn't mean that the decision to introduce a weight is a no-brainer. As you can see from our dice example above, including a weight means you need to keep track of the counts and adjust the probabilities after each round. For a six-sided die, this isn't such a big deal. But when you need to keep track of the counts for a list of hundreds or thousands of questions, across potentially millions of users, and update those counts (and associated probability distributions) whenever a round is played, this increases the complexity of your game by a not-insignificant amount. Whether introducing a weight is actually simpler than just increasing the size of the pool is something that game developers will have to decide for themselves.

2.7 THE COMPLETIONIST'S DILEMMA

There's one more thing I'd like to point out. Throughout this discussion, we've been assuming that delaying repetition is the ultimate goal of these types of games. But not everyone would agree with this. If you're a completionist—if your goal is to try your hand at every drawing in *Draw Something*, or test the limits of *Family Feud*'s answer recognition on every question, or, more generally, see everything that the game has to offer—then you may not care about extending the life of the game. In fact, you might be fine with a system that draws without replacement, so that you see no repeats at all until you've seen everything. For you, the central question may be: How many rounds do I need to play before I can expect to have seen everything?

If you're drawing without replacement from a pool of N questions, that's easy: it'll take you N rounds to see all of the questions. If you're drawing with replacement, then it's a little trickier. Let's start with a concrete example.

Suppose there are $N = 6$ questions, and you've seen 5 of them. Then the probability that you'll see a new question on your next round is 1/6, and on average, you should expect to play 6 more rounds in order to complete your quest to see everything. Similarly, if you've seen 4 out of the 6 questions, the probability that you'll see a new one on your next round is 1/3, and on average you should expect to play 3 more rounds before one of the last two questions makes an appearance. Continuing in this way, the total number of rounds you should expect to play before you see everything is

$$6+3+2+1.5+1.2+1=\frac{6}{1}+\frac{6}{2}+\frac{6}{3}+\frac{6}{4}+\frac{6}{5}+\frac{6}{6}=14.7.$$

That's 6 rounds to get the last question, 3 rounds to get the fifth, 2 rounds to get the fourth, and so on.

In general, if you've already seen m questions, the probability that the next question will be new is $\frac{N-m}{N}$, since there are $N - m$ new questions remaining and N questions in all. This, in turn, means that you should expect to need to play $\frac{N}{N-m}$ additional rounds before seeing your next new question.[15] This means that the total number of rounds you should

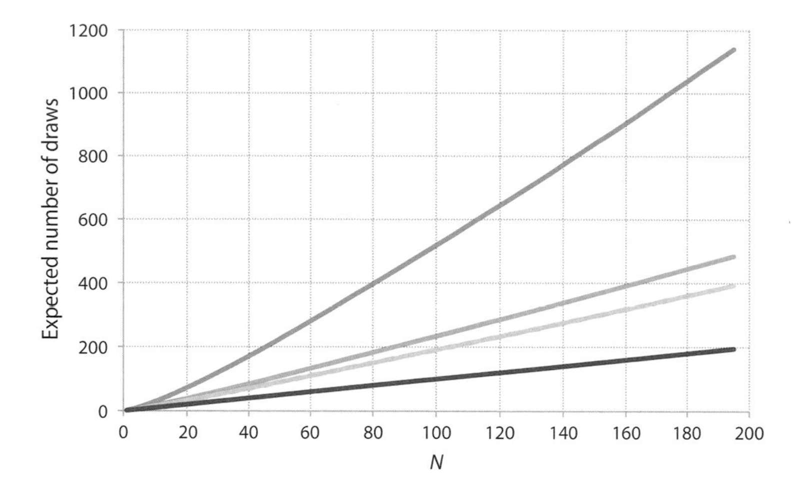

FIGURE 2.10. The expected number of draws to see everything when playing with replacement, with a weight of 1/2, with a weight of 1/3, and without replacement.

expect to play equals:

$$N + \frac{N}{2} + \frac{N}{3} + \ldots + \frac{N}{N} = N \sum_{m=1}^{N} \frac{1}{m}.$$

Clearly this expression is greater than N, whenever $N > 1$, which makes sense, since you should expect to play more rounds when drawing with replacement than when drawing without replacement.[16]

As before, between these two extremes we can also introduce a weight p to make repeats less likely. Also as before, the case $p = 1$ corresponds to drawing with replacement, and $p = 0$ corresponds to drawing without. These weight functions add complexity to the mathematics, but it's complexity that a computer can handle. Figure 2.10 shows a graph of the expected number of rounds you'd need to play in order to see everything versus the number of things, N.

Compared to Figure 2.9, there's a tradeoff here: decreasing the weight lowers the number of repeats (for a time), but in so doing exhausts the pool of questions more rapidly. If you're a completionist, this may not bother you. But for others, it might.

2.8 CLOSING REMARKS

When it comes to fighting repetition, there's only so much that game developers can do in the face of mathematical truth. Maybe introducing a weight function is the answer; maybe paying people to enlarge the pool you're drawing from is a better approach. Or maybe the best thing to do is to combine these two strategies, though completionists may disagree.

In the end, maybe the best we can do is shrug and try to enjoy the games that fall under the umbrella of this analysis as best we can. Remember that no matter the scope of the game, you're likely to see a duplicate fairly early on, but don't let that discourage you from continuing to play. Even when duplicates pop up with a fair degree of regularity, there's still plenty of content left to discover. And if you need to see everything the game has to offer, these types of games will likely keep you busy for quite some time.

2.9 ADDENDUM: IN SEARCH OF A MINIMAL k

In this section, we'll derive the inequality

$$k \geq \frac{1}{2} + \frac{\sqrt{8N \ln 2 + 1}}{2}.$$

If you're curious, read on. Fair warning: this part gets a bit technical, and if you change your mind you can safely skip the rest of this section.

The idea is to replace the product on the left side of the inequality

$$\left(1 - \frac{1}{N}\right) \times \left(1 - \frac{2}{N}\right) \times \ldots \times \left(1 - \frac{k-1}{N}\right) < \frac{1}{2}$$

with a sum, which is easier to deal with. We do this by making use of the following approximation: whenever x is a small number, the function $1 + x$ can be approximated fairly well by the function e^x (recall from your algebra 2 class that e is an irrational number whose decimal expansion begins with 2.71828...). In other words,

$$1 + x \approx e^x, \text{ for } x \approx 0.$$

This might seem like a completely random observation, but it's a natural thing to do if viewed through the lens of a much more powerful result in calculus.[17] For now, all we need is this approximation. (If you don't believe it, graph the two equations using your favorite piece of technology, and zoom in near zero.)

What does this approximation give us? Well, if N is big enough, then $-1/N$ is close to zero. So are $-2/N$ and $-3/N$. In fact, provided k isn't too large, every fraction from $-1/N$ to $-(k-1)/N$ is also close to zero. But if this is the case, then by using our approximation, we can write

$$\left(1-\frac{j}{N}\right) \approx e^{-j/N}$$

for any j between 1 and $k-1$. This, in turn, means that

$$\left(1-\frac{1}{N}\right) \times \left(1-\frac{2}{N}\right) \times \ldots \times \left(1-\frac{k-1}{N}\right) \approx e^{-\frac{1}{N}} \times e^{-\frac{2}{N}} \times \ldots \times e^{-\frac{k-1}{N}}$$
$$= e^{-\frac{1}{N}(1+2+\ldots+k-1)}.$$

This might not seem like much of an improvement, but the sum in the exponent can be simplified using the well-known formula[18]

$$1+2+\ \ldots\ +k-1=\frac{k\,(k-1)}{2}.$$

Pulling all of this together, we find that our original inequality can be approximated by the inequality

$$e^{-\frac{k(k-1)}{2N}} < \frac{1}{2}.$$

The inverse of the exponential function is the logarithm. If we take the natural logarithm of both sides, the inequality becomes

$$-\frac{k\,(k-1)}{2N} < \ln\frac{1}{2} = -\ln 2.$$

After rearranging this inequality, we get

$$k^2 - k - 2N\ln 2 > 0.$$

What's the smallest value of k for which this inequality is true? Figure 2.11 shows a graph of the quadratic on the left for a few different values of N.

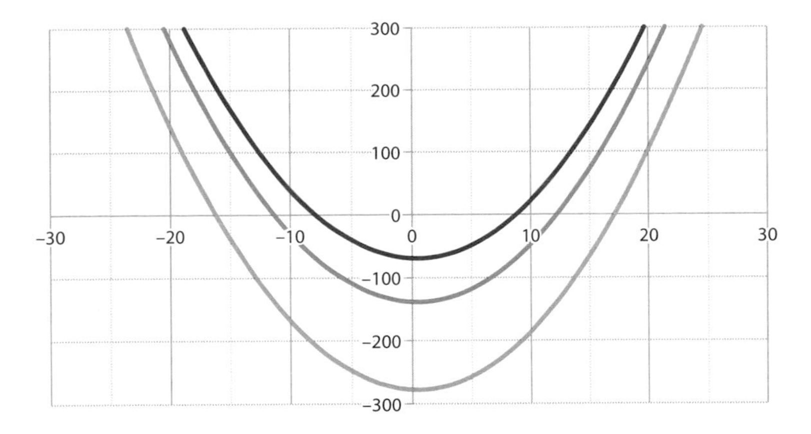

FIGURE 2.11. Graphs of the three quadratics in k corresponding to the cases $N = 50$, $N = 100$, and $N = 200$. In each case, the smallest value of k is the first integer larger than the rightmost root of the parabola.

Instead of considering an inequality, let's first take a look at the *equality*

$$k^2 - k - 2N \ln 2 = 0.$$

By the quadratic formula, the solutions to this equation are

$$\frac{1}{2} \pm \frac{\sqrt{8N \ln 2 + 1}}{2}.$$

As long as $N > 0$, the negative root is negative and the positive root is positive. Since k must be positive, it's the positive root that's relevant, and so in order for $k^2 - k - 2N \ln 2$ to be positive, we must have

$$k > \frac{1}{2} + \frac{\sqrt{8N \ln 2 + 1}}{2}.$$

3
Get Out the Voting System

3.1 EVERYBODY VOTES, BUT NOT FOR EVERYTHING

In early 2007, Nintendo released an app for the Wii called *Everybody Votes*. The idea was simple: players were given a set of prompts to choose from, and for each prompt, they could select one of two responses. For example, the prompt might say "I'd rather live in a house on . . . ," and players could choose between "a mountain" and "the beach." As a nice bonus, players could also predict which option would garner a majority of the vote before polls closed. After a few days, the poll would close and players could see how people all around the country (or world, depending on the poll) responded.

When the Wii U launched at the end of 2012, it served as a death sentence for many of the smaller apps that Nintendo had built for the Wii. *Everybody Votes* was no exception: the app shut down on June 28, 2013. The idea lives on, however; Nintendo released a game for the Wii U called *Splatoon* in 2015 that makes use of a similar idea. *Splatoon* features recurring events called "Splatfests," in which players choose one of two sides and then battle against the opposing side. At the end of the event, the winner is determined not only by which team played better but also by which side was the more popular one. Also as in *Everybody Votes*, the two sides in the Splatfest are represented by purple and green.

Two options. Two teams. In both games, players are never offered more than two choices. Sometimes the prompt naturally elicited only a couple of responses, but often this was not the case. And for those prompts that were more open-ended, the lack of options could sometimes feel stifling. In his review of *Everybody Votes,* IGN writer Mark Bozon had this very criticism:

> Even though the Everybody Votes channel delivers a bit of entertainment in short, short spurts, it's a bit too limited for our own tastes. For starters, each poll is made up of only two choices. As one of the starting polls on the Everybody Votes Channel the Wii asks users what the best valentines day [sic] gift is. The two choices: Roses, and chocolate. This makes sense, since those are the two top answers for the question, but jump over to a more elaborate question of "Where would you rather live?" and you might be a bit let down to see there are still only two answers: Beach, or Mountain. Nintendo: what's the deal-e-o? What if I want to live in a cave, in the wilderness, in a city, under the sea, or in a van down by the river? Can my choice not be heard too? It's a small gripe, but it's also a moronic move in our opinion. If you can't even ask what a person's favorite color is (or first-party Nintendo franchise, for that matter) without making it a "this or that" question, is it really going to bring back amazingly compelling data once the poll closes? (Bozon 2007)

At face value, this may seem like an entirely reasonable complaint, and it's easy to get swept up in Bozon's outrage. But if you think for a moment about the logistics of offering players multiple options, you will stumble upon a number of questions that the developers of *Everybody Votes* may simply not have wanted to deal with. For example, if you offer more than two options:

- What determines which option wins? (With two options, one of them must necessarily have a majority of the vote, but that won't be true with more than two options.)
- Do you let users express more detailed preferences? If so, how does that work? (For example, you may have strong feelings for living on the beach or in a city, but feel relatively indifferent about the other options. Should you have to choose between beach or city, or should you be able to tell the app that you like both?)
- If you let users express detailed preferences, how do you come up with a rule to determine a final ranked list of the choices?

As we'll see in this chapter, the answers to these questions are complicated. There are many different ways to implement a voting system, and voting isn't as simple as it seems. Certain voting systems can needlessly complicate an election; in fact, the winner of an election is sometimes determined more by the rules of the system than by voters' actual preferences!

Fortunately, games provide us with many examples of different voting systems. By examining how different games try to separate the best from the rest, we can come to a better understanding of what makes a voting system tick and why some voting systems are better than others.

3.2 PLURALITY VOTING: AN EXAMPLE

Let's imagine that Nintendo decided to release an updated version of *Everybody Votes*, with support for more than two choices in its polls. To make things more concrete, let's further imagine that their first poll question was the following: Which of the three major console manufacturers—Nintendo, Sony, or Microsoft—is the best? Best is a subjective term, of course, so we'll simply say that the best manufacturer is the one that voters prefer the most. How should we go about designing such a poll?

Arguably the simplest thing to do is use the *plurality* voting system. This is the most common way to hold an election in the United States, so the rules should sound familiar: every voter casts a ballot for his or her preferred console. After everyone has cast a ballot, the votes are counted, and the console with the most votes wins. For example, suppose that among 100 impassioned video game players, the votes broke down as follows:

Microsoft: 45 votes,
Sony: 40 votes,
Nintendo: 15 votes.

In this case, Microsoft would be declared the winner.

This is the way we elect many of our leaders in government, but that doesn't mean it's a perfect system. To the contrary, many criticisms have been levied against the plurality voting system. Some downsides are apparent even in this example. For instance, one could reasonably argue that none of these companies should win, since none of them has a majority of the votes. Maybe we should hold a runoff election in which Nintendo is scratched from the ballot, and voters are allowed to choose only between Microsoft and Sony. Indeed, maybe Sony is actually preferred by more voters than Microsoft, and Nintendo is acting as a sort of "spoiler" (similar to the effect that many people claim third-party candidates have in elections).

This gets to the heart of a more serious criticism: the plurality voting system does a poor job of capturing the full spectrum of a voter's preferences. For instance, if someone casts a vote for Nintendo, the ballot itself tells us nothing about that voter's feelings toward Microsoft or Sony. Maybe the voter loves Microsoft, but loves Nintendo a little bit more. Maybe he's a serious Nintendo fanboy and would rather die than own hardware from a competitor. The ballot itself provides a fairly limited amount of information. If the point of an election is to determine the "will of the people," it seems like plurality doesn't always do a great job capturing that will.

Of course, part of the problem also has to do with the number of candidates on the ballot. In the case of Nintendo vs. Sega, there are only two options to choose from. For an election between these two, absent a tie, one company is guaranteed to have a majority of the votes, and voter preferences are easier to infer from the ballot. In other words, in a two-party system, some of the flaws of the plurality system fade away. With more candidates, however, shortcomings of this system are harder to avoid.[1]

3.3 RANKED-CHOICE VOTING SYSTEMS AND ARROW'S IMPOSSIBILITY THEOREM

Plurality doesn't do a good job showing us how voters truly feel. So, maybe the simplest thing to do is ask voters for more information. What if, instead of asking for voters to identify their favorite candidate, we asked them to *rank* all of the candidates? This idea opens the door to a number of different election methods, collectively known as—you guessed it—*ranked-choice* voting systems. How do these stack up against the well-known plurality method?

TABLE 3.1. Breakdown of 100 Votes

1st place	2nd place	3rd place	Number of Votes
Microsoft	**Nintendo**	Sony	45
Nintendo	Sony	Microsoft	15
Sony	**Nintendo**	Microsoft	30
Sony	Microsoft	**Nintendo**	10

Let's return to our scenario from before and crunch the numbers to find the "best" console manufacturer using a ranked-choice voting system. There are many variants we could choose from, but to keep things manageable we'll examine only two:

Borda count. In this system, voters receive a ballot of n candidates and rank them from first to last place. Each first-place vote earns a candidate n points, each second-place vote earns a candidate $n - 1$ points, and so on, down to each last-place vote, which contributes one point. The candidate with the most points wins.

Instant runoff voting (IRV). As with the Borda count, voters receive a ballot of n candidates and rank them from first to last. Once the ballots are in, the candidate with the fewest first-place votes is eliminated, and those votes are distributed among the voters' second-place candidates. This process of eliminating the candidate with the fewest votes and redistributing those votes among the remaining candidates continues until one candidate has a majority of votes. (If this seems confusing to you, you're not alone! Voter confusion is frequently cited as a practical obstacle of this voting system.[2])

Let's see how these three methods compare to one another using the same set of candidates as before. Recall that with plurality, Microsoft received 45 votes, Sony received 40, and Nintendo received 15. Suppose that, with the introduction of the ranked ballots, those 100 votes broke down as in Table 3.1.

According to the table, all of our original Microsoft voters prefer Nintendo to Sony, and all of our Nintendo voters prefer Sony to

Microsoft. Among Sony fans, however, things are more complicated: 30 of them prefer Nintendo to Microsoft, while 10 prefer Microsoft to Nintendo.

Clearly, these ballots provide us with more information than we had before. Now we know which company each voter supports the most, as well as how voters feel about companies that didn't take the top spot.[3] But do these new ballots translate into the same election results as before, or will Microsoft be forced to cede its crown to a competitor?

The Borda count is the simpler ranked-choice method, so let's tackle it first. According to these ballots, Microsoft received 45 first-place votes, 10 second-place votes, and 45 third-place votes. This gives Microsoft a grand total of 200 points. Not too shabby.

What about Sony? It received 40 first-place votes, 15 second-place votes, and 45 third-place votes, for a total of 195 points. Once again, Microsoft comes out ahead, though just barely.

Nintendo, however, throws a wrench into our former narrative. This time around, Nintendo receives 15 first-place votes, 75 second-place votes, and only 10 third-place votes. This means that Nintendo winds up with 205 points. Just as Microsoft edged out Sony, so too does Nintendo edge out Microsoft for first place!

On the face of it, this may seem ridiculous—after all, Nintendo received the fewest first-place votes of any candidate. On the other hand, it also received the fewest third-place votes. In addition, a significant number of voters both preferred Nintendo to Microsoft and Nintendo to Sony. So maybe Nintendo is the rightful winner, after all.

What about instant runoff voting? Though tabulating votes can be somewhat complicated in practice,[4] in our current example it's fairly straightforward. Since Nintendo has the fewest first-place votes, it gets eliminated first. Of the 15 people who ranked Nintendo first, they all ranked Sony second, meaning that all 15 of those votes transfer to Sony. This gives Sony a majority total of 40 + 15 = 55 votes and the win over Microsoft.

So far, we've looked at three voting systems, and each one has chosen a different winner. At this point, the whole process of voting may seem like a pointless exercise. After all, if we can get an entirely different winner by changing the voting rules, what's the point of having a vote in the first place?

Then again, maybe we just haven't found the right voting system. And indeed, all three of the systems we've looked at so far (plurality, Borda count, IRV) suffer from the same serious flaw. It's easiest to understand this flaw first via an anecdote about former Columbia University philosophy professor Sidney Morgenbesser, who passed away in 2004. William Poundstone, author of *Gaming the Vote: Why Elections Aren't Fair (and What We Can Do about It)* tells the tale well:

> According to the story, Morgenbesser was in a New York diner ordering dessert. The waitress told him he had two choices, apple pie and blueberry pie. "Apple," Morgenbesser said.
>
> A few minutes later, the waitress came back and told him, oh yes, they also have cherry pie.
>
> "In that case," said Morgenbesser, "I'll have the blueberry" (Poundstone, 2008, p. 50).

If you think Morgenbesser sounds a little eccentric, you're not alone. Given the choice between apple and blueberry, it's clear he prefers apple. The addition of a new flavor (cherry) shouldn't influence his preference between apple and blueberry. It would make sense for him to stick with apple (meaning he preferred apple to both cherry and blueberry) or to switch to cherry (meaning he preferred cherry to apple, but apple to blueberry). But why on earth would the addition of cherry cause him to switch to blueberry?

The idea that a person's preferences between two choices, x and y, shouldn't depend on any other choices is called the *independence of irrelevant alternatives*, or IIA, criterion. In the anecdote above, it implies that Morgenbesser's preference between apple and blueberry pie shouldn't depend on whether or not an irrelevant alternative (such as cherry pie) exists. It seems reasonable that people's preferences shouldn't violate this condition.

Because IIA seems like a natural rule that (most) people live by, it stands to reason that a good voting system ought to follow this rule as well. Unfortunately, all of the systems we've looked at so far break this rule.

To see why, imagine that we conducted a poll using the plurality method between Microsoft and Sony. We already know that all the Nintendo voters prefer Sony to Microsoft, so Sony will win, 55 votes to 45. In other words, with Nintendo out of the picture, Sony wins over Microsoft. But with Nintendo in the picture, Microsoft is the winner! Here, Nintendo acts as the irrelevant alternative.

This same analysis works for the other voting systems, though the company playing the role of the irrelevant alternative changes. For the Borda count, Sony is the irrelevant alternative: when it is present, Nintendo wins, but when it is absent, Microsoft wins (by a vote of 55 to 45). For IRV, Microsoft is the irrelevant alternative: when it is present, Sony wins, but when it is absent, Nintendo wins (by a vote of 60 to 40). These irrelevant alternatives sometimes go by other names in voting circles: spoilers.

If we're looking for a voting system that satisfies IIA, we've struck out three times in a row. But let's not focus just on IIA; after all, there are probably a few other rules that we'd like our voting system to satisfy. Here are two more that seem entirely reasonable:

- Our voting system should satisfy *unanimity*: that is, if everyone ranks candidate A ahead of candidate B, then candidate A should come out ahead of candidate B in the final tally.
- Our voting system should satisfy *nondictatorship*. That is, loosely speaking, it shouldn't be a dictatorship. Slightly less loosely speaking, the outcome of the election shouldn't simply reflect the preferences of one person (a dictator) without taking into account the preferences of other voters.

Unfortunately, *no ranked-order voting system can satisfy these conditions and IIA simultaneously!* Put another way, any democracy using a ranked-order voting system that satisfies unanimity and the independence of irrelevant alternatives must be a dictatorship (which, in some sense, is the worst kind of democracy).

This result, known as Arrow's impossibility theorem, was first published by economist Kenneth Arrow in 1951 (though it should be said that the original formulation had slightly different assumptions; the version presented here is based on a modified version of the theorem published in 1963).[5] This unexpected result threw a wrench into the search for a "perfect" voting system, and to this day causes people to throw up their arms in despair when the topic of reforming our current system comes up. Since every system is flawed, they argue, why bother changing things at all?

3.4 AN ESCAPE FROM IMPOSSIBILITY?

Arrow's theorem highlights an important mathematical reality, and its conclusions are unavoidable based on its premises. However, the theorem makes one critical assumption that often gets overlooked: it assumes that we are dealing with a ranked-choice voting system.

If you've ever watched a movie on Netflix or reviewed a restaurant online, however, you're probably familiar with an example of a voting system that isn't ranked-choice. The key difference with these types of systems is that they let users *rate* things, instead of just *ranking* them.

For example, suppose voters in an election rated all candidates on a scale from 1 to 5 stars. Just like movie or product reviews, a score of 1 means that the voter strongly dislikes the candidate, while a score of 5 means that the voter strongly likes the candidate. A vote consists of a completed ballot with ratings on all candidates for which the voter has an opinion. When all the ballots are collected and tallied, the candidate with the highest score wins.[6]

This voting method is called *score voting* (for obvious reasons). It allows for ballots that are even more expressive than what you get out of a ranked-order method; this is because it allows voters to not only express preferences but also *degrees* of preference. With the previous methods, a voter could (at best) say that she preferred Sony to Microsoft, and Microsoft to Nintendo. With score voting, she can say that she loves Sony (5), but hates Microsoft (1) and Nintendo (1), or that she loves Sony (5) and Microsoft (5), but hates Nintendo (1), or loves all three or hates all three, or anything in between.

Intuitively, then, perhaps it's reasonable to expect that this added potential for expressiveness might translate into a better voting system. Of course, "better" is a highly subjective word, and advocates for voting reform have yet to agree on which system would provide the largest practical benefit over plurality. However, one thing is mathematically irrefutable: score voting satisfies IIA, unanimity, and nondictatorship. In other words, it escapes the clutches of Arrow's impossibility theorem.

TABLE 3.4. If 20 of the 45 People Who Originally Ranked Microsoft First Strategize and Vote for Nintendo over Microsoft, They Can Cause Sony to Lose!

1st place	*2nd place*	*3rd place*	*Number of Votes*
Microsoft	**Nintendo**	Sony	25
Nintendo	Microsoft	Sony	20
Nintendo	Sony	Microsoft	15
Sony	**Nintendo**	Microsoft	30
Sony	Microsoft	**Nintendo**	10

If score voting sounds like too much of a bother, there's an even simpler voting method known as *approval voting,* which is basically score voting with only two possible scores: 0 or 1. In other words, for any candidate on the ballot, you can either "approve" the candidate (1) or not (0). You can approve as many candidates as you like. Put another way, it's just like the well-known plurality system, but without the restriction that you can vote only for a single candidate.[7]

Robert Weber first coined the term *approval voting* (Weber 1977), and Steven Brams and Peter Fishburn wrote a paper on the voting method in 1978. Like score voting, approval voting is immune from the confines of Arrow's impossibility theorem. These two systems also enjoy some other nice properties that many other systems do not. For example, with score and approval voting, it never hurts to support your favorite candidate. This may seem like a trivial condition, but it's actually quite rare for a voting system to satisfy it. It's why third-party candidates fare so poorly in most U.S. elections: even if you honestly prefer a third-party candidate to everyone else who's running, you likely won't vote for that candidate since you know he or she has no shot of winning. In our current system, this is why voting for a third-party candidate is so commonly perceived as "throwing away your vote."

It's not hard to see how this condition fails with the ranked-order systems we've already considered. For example, in our hypothetical console election, 45 people ranked Microsoft first, Nintendo second, and Sony third. But suppose that 20 of those people, believing that Nintendo had a better shot of winning than Microsoft, strategically altered their ballots and flipped the order of Microsoft and Nintendo. This altered collection of ballots is summarized in Table 3.4.

With this tally, Nintendo now wins under IRV. So, by *not* supporting their favorite candidate, people who prefer Microsoft can prevent their least favorite candidate (Sony) from winning. Put another way, it's actually disadvantageous for those people to vote for their favorite candidate!

While neither approval nor score voting is totally immune from tactical voting, at the very least both are immune from this sort of behavior. With either system, you can fully support your favorite candidate, guilt free.

3.5 IS THERE A "BEST" SYSTEM?

We've now considered five different voting systems—plurality, IRV, Borda count, score voting, and approval voting—but it doesn't seem like we're any closer to determining a "best" system. And indeed, there is contentious debate among advocates for voting reform regarding which system we should use. While score voting is certainly highly expressive and satisfies the three criteria outlined in Arrow's impossibility theorem, it also demands much more from the voters, and like most systems isn't immune to strategic voting. Approval voting is simpler; in a sense, it's the closest to plurality, with the only difference being a change from "vote for one" to "vote for one or more." As a practical matter, then, approval voting would be the simplest system to implement in places where plurality is currently used. For these reasons and more, approval voting is the system advocated by the Center for Election Science, a nonprofit that is "dedicated to election-related scholarship."[8] At the same time, the nonprofit FairVote advocates for IRV as the best alternative to plurality.[9]

With so many different opinions, it's natural to wonder whether there's any way to objectively measure how these different systems stack up against one another. Mathematician Warren D. Smith explored this question in 2000. Smith was the first to compare different voting systems using the statistical concept of *Bayesian regret*. The concept itself was not new, but its application to voting systems was.

Roughly speaking, the Bayesian regret of an election system measures the "expected unavoidable human unhappiness" caused by the system.[10] For example, a voting system that continually elects unpopular candidates has a higher Bayesian regret than a system that tends to elect more preferred candidates.

In order to calculate the Bayesian regret for a voting system, one needs to know the following:

1. The number of candidates in the election;
2. The number of voters in the election;
3. The voting system used to elect the winner;
4. The amount of strategy voters use when casting their ballots (i.e., do the ballots align with voters' true feelings, or are they using strategy to cast dishonest ballots that they feel may help their preferred candidate?);
5. The distribution of voters' utilities.

Most of these inputs are fairly self-explanatory, but the last one requires a bit of unpacking. In order to measure how good or bad a voter will feel because of the outcome of an election, we need to use the concept of *utility* (known in some circles as *loss aversion*). For example, for a given voter, if one candidate wins, maybe that voter will gain 5 lifetime happiness units; if another candidate wins, maybe that voter will lose 20. This works as a reasonable proxy for the amount of satisfaction a voter feels regarding the outcome of an election. The last item in the list simply means that we know how each voter's utility will change based on the outcome of the election (that is, we know how each candidate affects each voter's utility).

With these five inputs, you can then calculate how much utility society gains from the outcome of the election, as well as the maximum possible utility society could have gained from each one of the candidates. The Bayesian regret of the election is simply the difference between these two values.

The first three items in the above list are fairly straightforward. Quantifying the last two might seem more difficult: after all, we have no way of knowing how much strategy voters might use in an election, and we certainly don't know voters' private feelings about every candidate. In fact, voters themselves often have no idea how their utility will be affected by each candidate, and so voter ignorance comes into play as well.

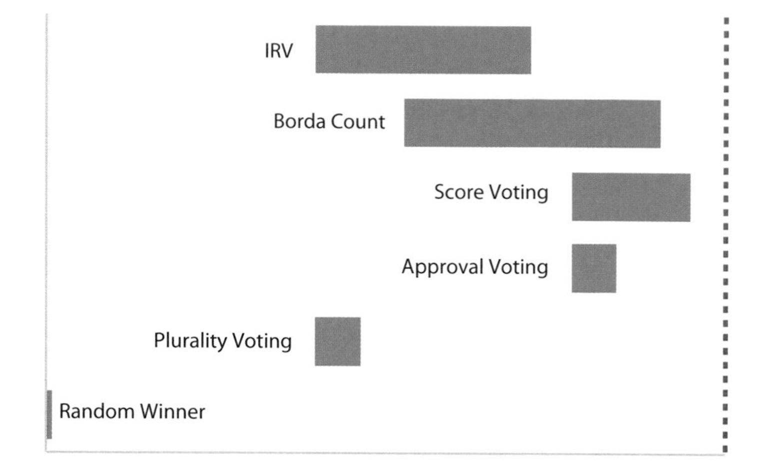

FIGURE 3.1. Comparison of Bayesian regret figures across the different voting systems we've considered. Values to the right correspond to lower Bayesian regret values; the dashed line on the right corresponds to an election with no regret. In this sense, bars to the right are "better" than bars to the left. Data from http://scorevoting.net/WarrenSmithPages/homepage/voFdata.

Smith accounted for these factors by running mock elections via computer simulation over hundreds of different combinations of utilities and levels of voter strategy. By averaging the results of millions of randomized simulated elections, Smith found ranges of Bayesian regret values for each of the voting systems we've considered so far. Figure 3.1 summarizes his results (adapted from the Center for Election Science, also featured in Poundstone (2008)).

We can infer quite a bit from Figure 3.1. First of all, regardless of the amount of strategy, plurality performs poorly. Also, the two alternative ranked-choice systems we considered both have quite a large range of Bayesian regret values, depending on how strategic voters choose to be. If voters are dishonest, IRV offers essentially no improvement over the status quo, while the Borda count offers a modest improvement.

Lastly, this chart shows yet another advantage to score and approval voting. Even if voters are strategic, these systems offer up a significant improvement over plurality. Between these two alternatives, score voting has a lower Bayesian regret than approval voting when voters act honestly, and they are the same when voters are strategic. This, along with the fact that approval voting is such a simple modification to our existing voting system, are major reasons why it is the preferred voting system of the Center for Election Science.

No voting system is perfect. But not all voting systems are created equal, either. The plurality voting system that people are most familiar with has a number of serious flaws. These flaws include, but are not limited to, the following:

1. Plurality creates the potential for spoilers;
2. It is one of the most inexpensive voting systems;
3. It falls under the conditions of Arrow's impossibility theorem.

When it comes to debating the merits of different game manufactures—or, you know, electing leaders of the free world—we'd be far better off if we used a different voting system. Bayesian regret figures suggest that approval voting and score voting in particular would yield tremendous gains. In the world of video game debates, that may not amount to much; but in the larger and thornier world of democracy, there's arguably nothing more important than ensuring that the "will of the people" is precisely that.

3.6 WHAT GAME DEVELOPERS KNOW THAT POLITICIANS DON'T

When it comes to the way we vote, the status quo is sorely lacking. Unfortunately, political change comes about slowly, and there's a great deal of work that needs to be done if the way we elect our leaders is to be modified on a large scale.

Game developers, however, are under no such constraints. Should the need arise, they can use whatever voting system they want. And the need does arise: because of the growth in social gaming, customization, and user-generated content, developers are routinely interested in soliciting the opinions of the people who play their games.

Let's take a quick look at two examples of how developers attempt to gather information on the opinions of their users, by way of the *Assassin's Creed* and *LittleBigPlanet* series.

In the *Assassin's Creed* games, players take on the roles of assassins during different periods in world history, including the Third Crusade, the Renaissance, and the American Revolution. The main campaign of each game has players progress by completing missions that advance the story. Some of these missions feature action-packed assassinations, while others are tension-filled stealth sequences. But not all missions are created equal. In the earlier games especially, some missions felt like

FIGURE 3.2. An example of the (optional) rating mechanic that users are presented with after successful completion of a mission in *Assassin's Creed IV*. Players can rate each mission from one to five stars (or opt out of voting altogether). © Ubisoft Entertainment. All Rights Reserved. *Assassin's Creed*, Ubisoft, and the Ubisoft logo are trademarks of Ubisoft Entertainment in the U.S. and/or other countries.

filler between more elaborate set pieces. The first game, for example, featured a number of missions in which the goal was to sit on a bench and listen to people have a conversation. Thrilling stuff.

At first, the simplest way for unhappy players to complain about boring missions was on Internet message boards. While this act may have brought about a certain degree of catharsis, it was also not necessarily the best way to facilitate change.

In order to tighten the feedback loop between player and developer, *Assassin's Creed IV: Black Flag*, which was released in the fall of 2013, introduced a new mechanic. After every mission, players were given the option to rate that mission, on a scale from 1 to 5 stars (Figure 3.2). In other words, for the first time players were given the opportunity to vote for their favorite (and least favorite) missions.

It shouldn't be surprising that the developers decided to implement score voting. After all, it's the most expressive voting system we've encountered, and absent an actual comment field it's probably the best way to gather feedback on the popularity of missions. (The fact that most users are already familiar with score voting from ratings on sites like Amazon, Netflix, and Yelp probably doesn't hurt either.)

Ubisoft, the company behind the *Assassin's Creed* games, has gone on record as saying that these user-generated ratings will be used to make improvements to future games (and so far, they've stuck to their word). Around the time of *Assassin's Creed IV*'s release, Ubisoft User Research Project Manager Jonathan Dankoff offered up the following statement:

> Ubisoft has been using data tracking to improve our titles for a while, but this is the first time we will be able to match gameplay metrics to player appreciation in order to dig even deeper into player behavior. ... The combination of the two data sources will give us incredible insight into how players are interacting with our game and guide future development teams to create missions that appeal to our players even more.[11]

Assassin's Creed isn't the first series to have embraced the idea of rating in-game content. The *LittleBigPlanet* series, for example, has used ratings for years, though for slightly different reasons and using slightly different rules. In *LittleBigPlanet*, customization is key: not only can players customize the look of their Sackboys and decorate the levels they play with stickers and other colorful accoutrements, but they also can create entire levels for people from around the world to enjoy. With millions of levels to choose from, separating the good from the bad becomes a fairly important task. By letting players rate the levels they play, the developers can also curate the user-generated levels, while budding game designers can receive feedback on what works and what doesn't.

As for the system used to rate levels, *LittleBigPlanet* has taken an interesting trajectory. When the first game in the series was released in 2008, the rating system was the same as the one used in *Assassin's Creed*: after completing a level, players could optionally rate it on a scale from one to five stars. When browsing the levels, players could then see each level's average star rating and use that to gauge whether the level was worth playing (Figure 3.3). Similarly, players could also see how many other people had played the level and how many had "hearted" it. "Hearting" a level added it to a personal list of your favorite levels and could therefore be used as a sort of bookmark feature so that you could later return to levels that you thought sounded interesting.

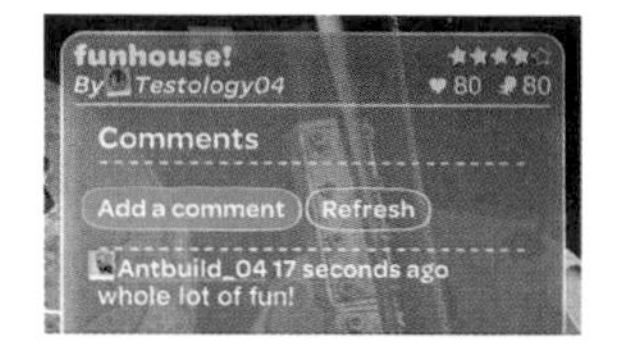

FIGURE 3.3. Example of the star ratings (along with commenting features) available when the original *LittleBigPlanet* launched in 2008. The heart icon represents the number of times a level has been "favorited," while the Sackboy icon represents the number of people who have played the level. © Sony

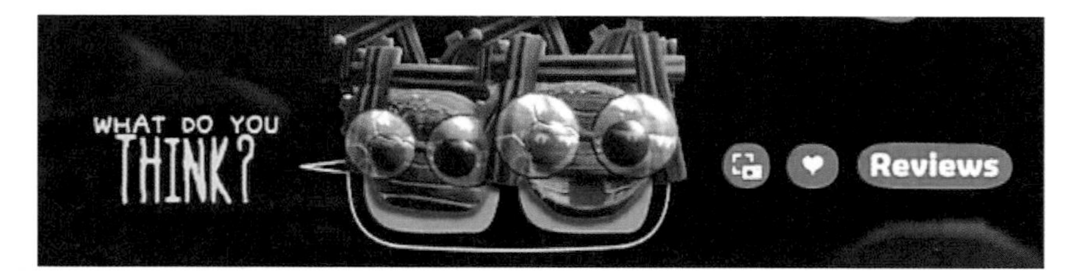

FIGURE 3.4. An example of a sad Sackboy (left) and a happy Sackboy (right). Screenshot from *LittleBigPlanet 2*. © Sony

By the time that *LittleBigPlanet 2* released in 2011, the voting system had changed (Figure 3.4). Instead of star ratings, users were prompted to give the level a happy face ("yay!") or a frowny face ("boo!"). Of course, this was still a form of score voting, just with a simplified scale: instead of rating levels from 1 to five stars, players could rate levels with either a frowny face (−1), a happy face (+1), or no face at all (0).

With this change in the scoring came a change in how the scores were reported. Instead of showing an average, as is typically done with star ratings, the level information simply began including the number of happy and sad ratings, as shown in Figure 3.5.

Just as with the star ratings, though, the happy/sad dichotomy was not long for this world. Eventually the sad face was retired, so that players could no longer rate a level negatively; they could give it a happy face or nothing at all.

In effect, then, the rating system in *LittleBigPlanet* changed over time from a traditional score voting system, to a more simplified score voting system, and finally to the most simplified version of score voting: approval voting. There are two effects to help explain these transitions. First, as a practical matter, there are a lot of young children who play

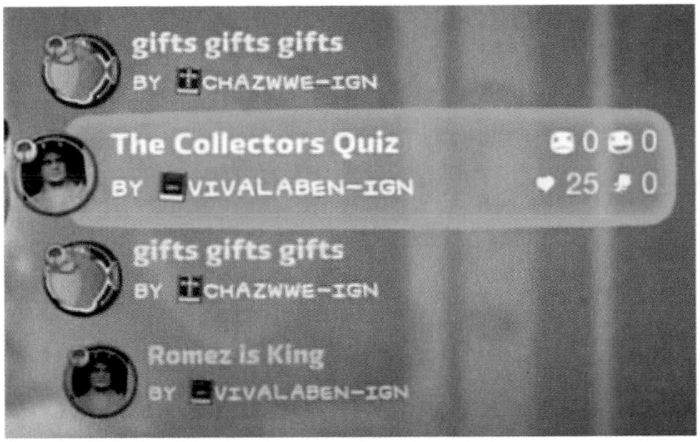

FIGURE 3.5. An example of the layout in *LittleBigPlanet 2*, for a level with zero playthroughs. Screenshot from *LittleBigPlanet 2*. © Sony

and create levels in *LittleBigPlanet*; for these players, the negative reviews can be harder to take, and there's also a risk that players hand out negative reviews without much justification. When the developers removed the option to "boo" a level in *LittleBigPlanet 2*, they also released the following statement explaining their decision:

> To help encourage creativity, we have removed the Boos from the Level Rating System. You can still Yay a level to say that you really liked it, but we found that the Boo ratings were disheartening at best, and could be used to cause grief among the community (Isbell 2012).

Another benefit to using a simplified system like approval voting is that it encourages more people to vote. Rating a level on a scale from 1 to 5 can be a bit of a chore; deciding what separates a four-star level from a three-star level, for example, may be more trouble than it's worth. As a result, a player may simply opt not to rate the level at all. By including a simple "like" feature, voting becomes easier for the player. The hope is that this translates into more votes, since the more votes the *LittleBigPlanet* levels have, the easier it will be for the curation process to do its job.

It should be noted that the changes didn't stop there. Perhaps realizing that without an option to vote negatively, "liking" and "hearting" were effectively the same, developers eventually merged the two.

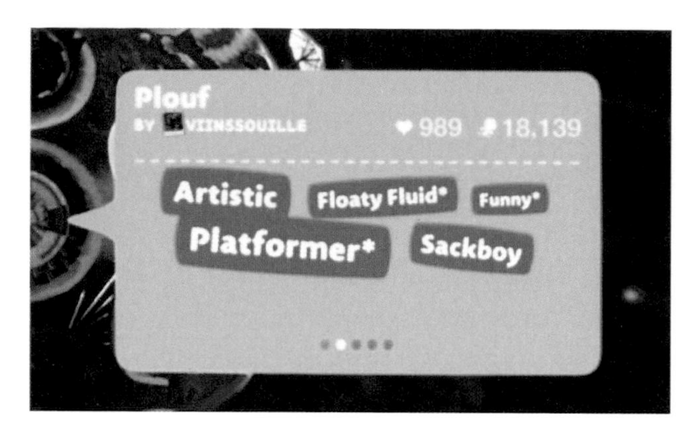

FIGURE 3.6. Level rating in *LittleBigPlanet 3*. In the upper right, you can see counters for hearts and plays, but not likes or dislikes. © Sony

As of 2016, you don't see happy faces at all when looking at level summaries; instead, "liking" and "hearting" are now basically synonymous (Figure 3.6).[12]

3.7 THE BEST OF THE REST

Of course, the purpose of curation in *LittleBigPlanet* is different than in our earlier hypothetical example of console manufacturers. In the latter case, we were trying to determine a *winner* among a pool of candidates. But for developers and players of games like *LittleBigPlanet* and *Assassin's Creed*, the purpose of the ratings is less focused on finding the best that the game has to offer and more focused on weeding out the worst. With so many levels at a player's fingertips, it's critical that *LittleBigPlanet* be able to separate the levels that players like from the levels that they don't.[13]

Fortunately, *LittleBigPlanet* does a reasonably good job of separating the wheat from the chaff. When you're looking for a level to play, there are a few ways to find good levels quickly. For example, the developers have created a pool of user-created levels that they are particularly fond of. This approach is certainly a dictatorship, but at least it's a benevolent one.

Another way to sort levels is based on "liking"/"hearting" functionality, or by the number of plays. Most-played is a fairly straightforward

TABLE 3.5. Highest-Rated *LittleBigPlanet 2* Levels in January 2015

Level Name	*Author*	*Likes*	*Hearts*	*Plays*
NEW! Five Nights at Freddy's 2 [2-4 Players]	seabreez15	1,213	1,565	6,329
Night watch at freddy fazbear's pizza 2 part 1	x-psycho6-x	1,020	1,204	5,140
five nights at freddy (hide and seek 2-4 new version	gab2222233	976	1,008	5,843
Night watch at freddy fazbear's pizza 2 part 2/final	x-psycho6-x	943	1,010	4,736
Marionette's Story	boz_tacticalnuke	729	706	4,526
Haha-Jima Island RPG/Hangout	heartbreaker215	511	874	2,961
Minecraft Mortal Combat	rembrant2005	459	408	2,649
five nights at freddy2 rpg new version 1.2	gab2222233	411	444	2,756
MORTALKOMPAT4EVER [SUBWAY]	ayumi_death	384	342	2,448
five night at freddy 2	alandeleon	384	309	2,769
MARIO WORLD (Platformer)	lacantalou	368	400	2,147
mortal combat marvel	aman701	356	261	1,697
Shark Survival	insalatamista456	345	464	2,455
T.U.1's Profile Level (Update 1.15)	the-unknown-one	341	301	1,888
five night st freddys 2 costumes update 1.11	minimoocifer2	341	343	2,102
Beta File: "DECEPTION" extracted	the-unknown-one	314	348	1,500
<Slapsters: Lighting Obsidian Edition>	thetslap-sters	312	426	1,808
THE FUN SKATE	insalatamista-2	302	407	1,946
Colorful Bomb Survival Yay/ Like for Heart	zskywalker78z	301	226	1,813
Bomb Survival...IN SPACE!!!	mmm_check_please	265	256	1,305

metric, but the best-rated levels merit further consideration. For example, Table 3.5 provides a list of the top 20 highest-rated levels (i.e., levels with the most likes) in *LittleBigPlanet 2* for January 2015.

As you can see, the levels are sorted by number of likes, with the highest totals up top and the lowest below. But is this really the best way to determine the ranking? Sure, the top-ranked level on this list has the largest number of likes, but it also has the largest number of plays. It doesn't seem right that we should ignore the play total; after all, if 1 million people played a level and 600 people "liked" it, that would put this level at the top of the list even though nearly everyone who played it *didn't* "like" it.

To put it another way, consider the following scenario. Suppose you're presented with two levels: one has 600 likes and 1 million plays, while another has 500 likes and only 500 plays. Clearly the first level has more likes, but it's debatable whether it should be considered the best-rated, since less than a tenth of one percent of people who have played the first level liked it. In comparison, 100% of people who played the second level liked it.

Maybe, then, we should be ranking based not on the absolute number of likes but on the *percentage* of likes. If we did that, Table 3.6 shows how the top 20 would change.

The rightmost column in Table 3.6 shows each level's new ranking, along with the change in rankings compared with those of our earlier approach. As you can see, the rankings change quite a bit. For instance, the top three levels based on percentage of likes were in the bottom half of the rankings when we considered absolute number of likes.

While using percentages may seem more reasonable, this isn't perfect either. To see why, suppose you wanted to compare a level with 1,000 likes and 2,000 plays to one with 55 likes and 100 plays. Ranking by number of likes, the first level would come out ahead. By our second approach, the second level would take the crown.

However, the 55% statistic should be taken with a grain of salt, since the first level has 20 times as much information (i.e., plays) as the second. Percentages are able to paint a more accurate picture only if you've got enough paint, that is, if enough people have played the level. What we really want, then, is to balance the percentage of likes against the absolute number of plays. Percentages based on a high number of plays should be taken more seriously than percentages based on a small number of plays.

What's the best way to accomplish this goal? One approach revolves around the Wilson score confidence interval, named after American mathematician Edwin Wilson.

TABLE 3.6. Sorting by Like Percentage

Level Name	*Likes*	*Plays*	*% Likes*	*Rank*
mortal combat marvel	356	1,697	21.0%	1 (+11)
Beta File: "DECEPTION" extracted	314	1,500	20.9%	2 (+14)
Bomb Survival...IN SPACE!!!	265	1,305	20.3%	3 (+17)
Night watch at freddy fazbear's pizza 2 part 2/final	943	4,736	19.9%	4 (0)
Night watch at freddy fazbear's pizza 2 part 1	1,020	5,140	19.8%	5 (−3)
NEW! Five Nights at Freddy's 2 [2-4 Players]	1,213	6,329	19.2%	6 (−5)
T.U.1's Profile Level (Update 1.15)	341	1,888	18.1%	7 (+7)
Minecraft Mortal Combat	459	2,649	17.3%	8 (−1)
Haha-Jima Island RPG/Hangout	511	2,961	17.3	9 (−3)
<Slapsters: Lighting Obsidian Edition>	312	1,808	17.3%	10 (+7)
MARIO WORLD (Platformer)	368	2,147	17.1%	11 (0)
five nights at freddy (hide and seek 2-4 new version	976	5,843	16.7%	12 (−9)
Colorful Bomb Survival Yay/Like for Heart	301	1,813	16.6%	13 (+6)
five night st freddys 2 costumes update 1.11	341	2,102	16.2%	14 (0)
Marionette's Story	729	4,526	16.1%	15 (−10)
MORTALKOMPAT4EVER [SUBWAY]	384	2,448	15.7%	16 (−7)
THE FUN SKATE	302	1,946	15.5%	17 (+1)
five nights at freddy2 rpg new version 1.2	411	2,756	14.9%	18 (−10)
Shark Survival	345	2,455	14.1%	19 (−6)
five night at freddy 2	384	2,769	13.9%	20 (−11)

The derivation of this score isn't particularly relevant, but the curious reader can find the details in the addendum at the end of this chapter. For now, here's the formula: for a level with n plays and a like percentage of p_0, the score is given by

$$s = \frac{p_0 + \frac{1.96^2}{2n} - 1.96\sqrt{\frac{p_0(1-p_0)}{n} + \frac{1.96^2}{4n^2}}}{\left(1 + \frac{1.96^2}{n}\right)}.$$

This formula may not be easy to parse, but it uses the information we have about likes and plays for a given level to make a statistical prediction of what the percentage of likes *would be* if every *LittleBigPlanet* player played that level. In other words, there is a 97.5% probability that the "true" fraction of likes (if you had complete information about *LittleBigPlanet* player preferences) would be at least as big as s.

Returning to our earlier hypothetical example, we see that a level with 1,000 likes and 2,000 plays has n = 2,000, p_0 = 0.5, and $s \approx 0.479$. On the other hand, a level with 100 likes and 55 plays has n = 100, p_0 = 0.55, and $s \approx 0.474$. In other words, even though the first level has a smaller percentage of likes, this percentage should carry more weight because so many more people have played (and rated) the level.

Put another way, suppose every *LittleBigPlanet* owner played these two levels and decided whether or not to like each level. Based on the information we have, there's a 97.5% probability that the first level would have at least a 47.9% like rate, while there's a 97.5% probability that the second level would have at least a 47.4% like rate. Since the first rate is higher, we rank the first level higher (though it's a relatively even matchup).

What happens to our top-twenty list of levels using the Wilson score? Table 3.7 summarizes the results. Total likes, percent likes, and the Wilson score are shown; the changes in rankings are relative to the total likes ranking.

The Wilson score also produces a noticeable reordering of the original rankings, though it's much more closely aligned to the ranking based on percentages. In fact, between these two rankings, no level shifts up or down by more than two places.

When we replace a simple 0–1 scale with something more expressive (the star ratings *LittleBigPlanet* used to use, for instance), level rankings become even more complex. But the same principles apply: there's still a need to balance the average rating against the amount of information present. This is something many websites still haven't perfected; occasionally on Amazon.com, for example, an item with a small number of ratings will be placed higher than one with a large number of ratings, even if their average ratings are comparable (Figure 3.7).

Of course, more complexity isn't necessarily better. On the other end of the spectrum, the algorithm behind Apple's App Store rankings isn't known to developers, though there are some conjectures about how it works. What's clear is that Apple has a large body of data from which to compile its rankings: not just user ratings, but also revenue, number of

TABLE 3.7. Comparing Like Percentage to the Wilson Score Confidence Interval

Level Name	*Rank, Likes*	*Rank, % Likes*	*Rank, Wilson*
mortal combat marvel	12	1 (+11)	1 (+11)
Beta File: “DECEPTION” extracted	16	2 (+14)	2 (+14)
Bomb Survival...IN SPACE!!!	20	3 (+17)	5 (+15)
Night watch at freddy fazbear’s pizza 2 part 2/final	4	4 (0)	3 (+1)
Night watch at freddy fazbear’s pizza 2 part 1	2	5 (−3)	4 (−2)
NEW! Five Nights at Freddy’s 2 [2-4 Players]	1	6 (−5)	6 (−5)
T.U.1’s Profile Level (Update 1.15)	14	7 (+7)	7 (+7)
Minecraft Mortal Combat	7	8 (−1)	9 (−2)
Haha-Jima Island RPG/Hangout	6	9 (−3)	8 (−2)
<Slapsters: Lighting Obsidian Edition>	17	10 (+7)	12 (+5)
MARIO WORLD (Platformer)	11	11 (0)	11 (0)
five nights at freddy (hide and seek 2-4 new version	3	12 (−9)	10 (−7)
Colorful Bomb Survival Yay/Like for Heart	19	13 (+6)	14 (+5)
five night st freddys 2 costumes update 1.11	15	14 (0)	15 (−1)
Marionette’s Story	5	15 (−10)	13 (−8)
MORTALKOMPAT4EVER [SUBWAY]	9	16 (−7)	16 (−7)
THE FUN SKATE	18	17 (+1)	17 (+1)
five nights at freddy2 rpg new version 1.2	8	18 (−10)	18 (−10)
Shark Survival	13	19 (−6)	19 (−6)
five night at freddy 2	10	20 (−11)	20 (−11)

times someone opens the app, average time spent in the app, number of times the app has been installed in the past few hours/days/months, and so on. And even though the algorithm isn’t explicitly known, developers have occasionally used their best guess about how the rankings might be compiled to try and give their apps an edge.

One example is the game *Flappy Bird*, which soared to the top of the App store in 2013 despite a lack of marketing on behalf of its developer, Dong Nguyen, and a decidedly mixed reception to the game in the media.[14] The game itself is quite simple. There’s a bird on the screen who is gradually falling toward the ground under the force of gravity. When you tap the screen, the bird flaps its wings and moves up toward the sky. Your goal is to navigate the bird through an infinitely long

FIGURE 3.7. There are also strange situations like this, in which one item will be wedged between two items that have the same number of ratings and the same average rating.

sequence of obstacles; the longer the bird survives, the higher your score.

After the game became a viral sensation, criticism began to grow claiming that Nguyen had essentially gamed Apple's algorithm in order to get more visibility in the App Store.[15] Certainly he had a financial incentive to do so; according to Nguyen, at the height of its popularity the game was earning him upwards of $50,000 a day.[16]

Regardless of how the rankings are actually determined, the point is that a straight count of positive ratings, like a straight tally of first-place votes, is rarely sufficient. Just as the determination of a winner requires careful consideration, so too does the subsequent ranking of everyone else.[17]

3.8 CLOSING REMARKS

Winning and losing are a part of life, but how we select the winners and losers can be just as important as the winners and losers themselves. As we've seen, this has implications far beyond the next video game: selecting winners and losers is an important part of almost all online commerce, and the problem of selecting one winner from a pool of many is of fundamental importance in our present-day democracy.

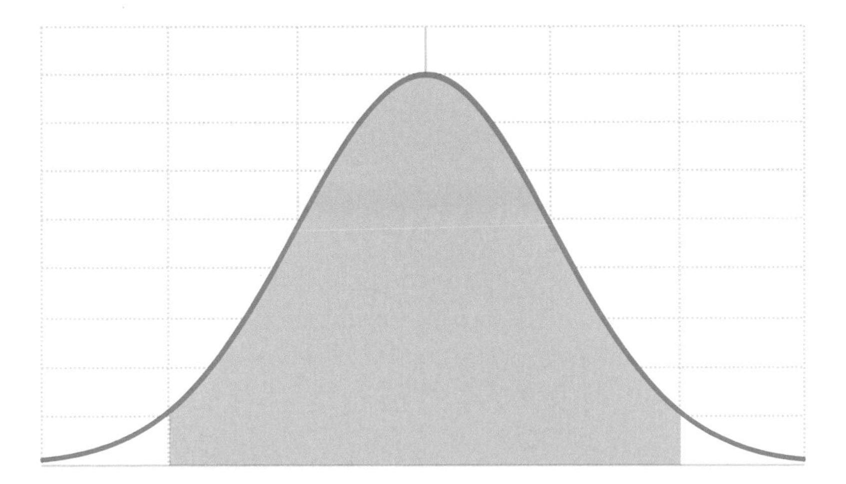

FIGURE 3.8. A standard normal curve, with 95% of its area shaded blue.

Because game designers are free of the shackles imposed in many other real-world examples of ratings and rankings, games have occasionally stumbled upon improvements not seen in other areas. Even in some of our favorite games, however, the best solution isn't always the implemented solution. An added sprinkle of mathematical thinking could, at the very least, help make future games even better. Outside of gaming, the implications could be even more profound.

3.9 ADDENDUM: THE WILSON SCORE CONFIDENCE INTERVAL

For simplicity, to derive the Wilson score formula, we'll stay within the context of *LittleBigPlanet*. To understand how the Wilson score confidence interval works, suppose that each level has some "true" probability p that a player will like it. We don't know what p is, of course, but we can estimate it by waiting for some number of people (say n of them) to play the level, and calculating the proportion of those players who liked the level. Let's let p_0 stand for this proportion.

When np and $np(1 - p)$ aren't too small, then by the central limit theorem, the distribution of p can be reasonably approximated by a normal distribution. In particular, this means that the probability that p_0 is a particularly bad estimate for p can be estimated using the area under a standard normal curve.

To put it another way, 95% of the area under a standard normal curve lies within 1.96 standard deviations of the mean (Figure 3.8). This means that the probability that p_o lies within 1.96 standard deviations of p is approximately given by

$$P\left(p - 1.96\sigma < p_o < p + 1.96\sigma\right),$$

where $\sigma^2 = p(1 - p)/n$ is the variance in the proportion of likes.[18] Put another way, for a 95% probability that the observed probability p_o won't be too far off the mark, we need

$$P\left(-1.96 < \frac{p_o - p}{\sigma} < 1.96\right) = 95\%.$$

Unfortunately, we don't know what p is! The best we can do is look at the endpoints in our interval

$$-1.96 < \frac{p_o - p}{\sigma} < 1.96$$

and try to solve for p in these extreme cases. After squaring and rearranging terms, we can rewrite this inequality as

$$(p_o - p)^2 < (1.96\sigma)^2 = 1.96^2 \frac{p(1-p)}{n}.$$

Therefore, at either end of the interval, we have

$$(p_o - p)^2 = 1.96^2 \frac{p(1-p)}{n}.$$

But this is just a quadratic equation in p. By expanding the left side and collecting like terms, we can rewrite the above equality as

$$p^2\left(1 + \frac{1.96^2}{n}\right) - p\left(2p_o + \frac{1.96^2}{n}\right) + p_o^2 = 0.$$

The quadratic formula then gives us the following two solutions for p:

$$p = \frac{p_0 + \frac{1.96^2}{2n} \pm 1.96\sqrt{\frac{p_0(1-p_0)}{n} + \frac{1.96^2}{4n^2}}}{\left(1 + \frac{1.96^2}{n}\right)}.$$

From this we can say that there's around a 97.5% chance that the true probability p is at least as large as the smaller of these two solutions (i.e., the one with the negative root). This is where the Wilson score comes from.[19]

4

Knowing the Score

4.1 RANKING PLAYERS

In the last chapter we saw how different voting systems can affect the outcome of an election. We also looked at the problems that arise when you try to organize a collection of levels in a game from best to worst.

But ranking games isn't the only way to get into a debate with a gaming aficionado. Instead of debating the merits of the games themselves, you can also debate the skills of those who play them. Unlike arguments about games, though, there's often a natural and impartial arbiter when it comes time to settle debates about who is the better player: the *high score*.

Popular in arcade games of the late 1970s and 1980s, high scores began to fade as home consoles became more popular. But with the rise of the Internet, along with social and mobile gaming, the competition to claim the highest score has seen a resurgence. The main difference now is that the competition, rather than being restricted to a single community of local arcade buffs, is now global.

In order to have a fighting chance to make it onto the scoreboards, there are at least three things a player needs to have. First is skill in the game itself; if you're bad at the game, you're not going to earn a high score. Second is luck; while skill is necessary to achieve a high score, it typically isn't sufficient, and even the best players will sometimes have a bad game.

The last thing a player needs is a deep knowledge of the game's scoring mechanics. You might be awesome at *Pac-Man* and have a really good run one day, but if you don't know how many points you earn from eating ghosts, pellets, and different fruits, you're not going to effectively modify your playing strategy in order to maximize your score.

High scores offer a relatively easy way to rate your performance in a game. As an added bonus, there's often some really great math to be gleaned from looking at how scores are calculated in certain games. In this chapter, we're going to explore some games with particularly interesting scoring rules. We'll then return to the phenomenon of high scores in general and try to predict how high humankind might reach in the quest for the highest score.

4.2 ORISINAL ORIGINAL

In my college days, I became somewhat of a connoisseur of free online games. (This was due, in large part, to a somewhat overzealous procrastination phase that I have mostly grown out of. Mostly.) As with any other type of content, there's a lot of garbage in the realm of free online games, but every once in a while I stumbled upon a treasure trove that would keep me otherwise occupied for a few hours, if not a few days.

One such repository goes by the name of Orisinal. Created by Ferry Halim, Orisinal is a collection of around 60 Flash games that can be played online for free.[1] Halim has been building these games since 2000, and by now a certain aesthetic has emerged: most of the games favor simple mechanics, a beautiful design, relaxing music, and adorable animals.

Most importantly for the present discussion, the games all feature some sort of scoring system. The high scores aren't saved globally, but they are saved locally; in other words, each game keeps a record of your personal best score in case you want to come back later and compete against yourself.

Some of these scoring systems are more interesting than others. One of the best is packaged in a charming little game called *Sunny Day Sky*. The rules are displayed every time you fire up the game, and are shown in Figure 4.1.

The rules probably don't make a whole lot of sense out of context, but once you start playing the game, it all comes together. You take on the role of a small bear who, like Mary Poppins, can use an umbrella to fly. Opening the umbrella shoots you up; closing it sends you plummeting to the Earth, but reopening it will send you skyward once more.

FIGURE 4.1. How to play *Sunny Day Sky*. Reproduced with permission from Ferry Halim.

There's more to this story, though: this unfortunate bear happens to find himself in the middle of a busy highway. Because of this, whenever he lands, he needs to do so on top of a moving vehicle. If he does, he can take off again and the game continues; but if he fails to stick the landing, he's run off the road and the game ends.

As with any other game, the best way to understand *Sunny Day Sky* is to play it. Alternatively, Figure 4.2 shows some screenshots that should give you a decent idea of how the game flows.

Given these rules, then, how exactly does the scoring work? It turns out that the answer depends on where you play the game. Whether accidental or intentional, the scoring rules are different between the browser version of the game and the mobile version that was released for iPhones in 2010. Since the rules in the mobile version are slightly more mathematically interesting, they're the ones we'll focus on here.[2]

You score points by flying over cars. The more cars you can fly over before landing, the more points you score. You don't score any points for passing your first car; after passing your second car you earn 10 points; after your third, you earn another 20, and so on. Table 4.1 summarizes the pattern.

This isn't the only way to score points, however. You also score whenever you make a successful landing. This is called a "nice landing" bonus, and it too grows with the number of cars you fly past.

This bonus grows in a similar way to the base score. If you land on the first car you pass, your bonus is 10 points. If you land on the second car you pass, your bonus increases to 10 + 20 = 30 points. After the third car, it increases to 10 + 20 + 30 = 60 points. Therefore, assuming you land successfully, your total score for your flight equals your base score plus

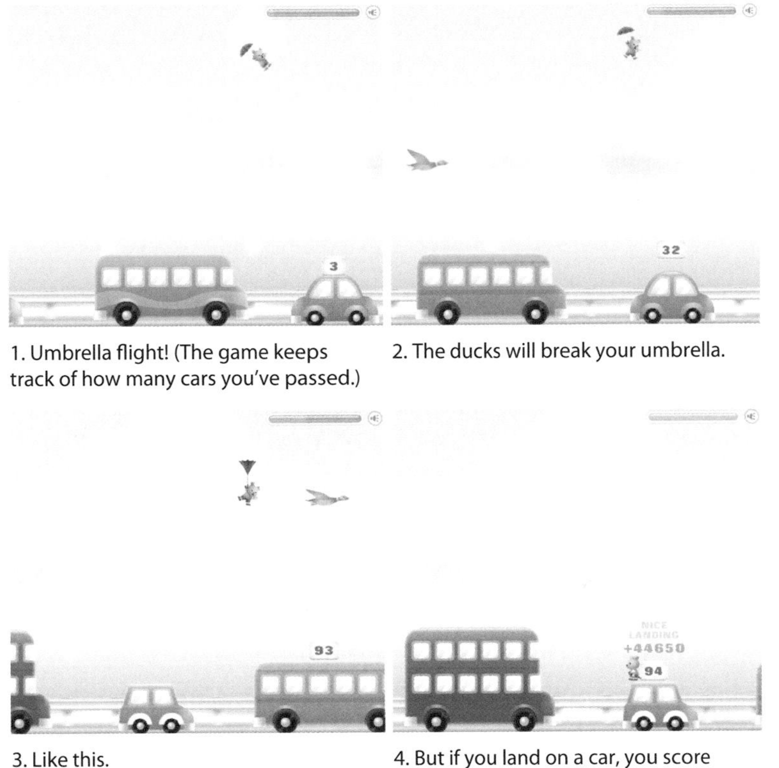

1. Umbrella flight! (The game keeps track of how many cars you've passed.)

2. The ducks will break your umbrella.

3. Like this.

4. But if you land on a car, you score a lot of points!

FIGURE 4.2. The mechanics of *Sunny Day Sky*. Reproduced with permission from Ferry Halim.

TABLE 4.1. Floating over Your Fourth Consecutive Car Earns You 30 Points and Brings Your Total Score up to $10 + 20 + 30 = 60$ Points

After passing car number...	*1*	*2*	*3*	*4*
You score an additional...	0 points	10 points	20 points	30 points
Your cumulative score is...	0 points	10 points	30 points	60 points

your "nice landing" bonus. Table 4.2 shows the scores in a small number of cases.

Your score is always divisible by 10, so let's clean up the notation by dividing all the scores by 10. We're left with the results in Table 4.3.

By now, we can see some patterns emerging. First, using this simplified scoring rubric, flying over your n^{th} consecutive car increases

TABLE 4.2. Base Score, Bonus, and Total

Landing on car number...	*1*	*2*	*3*	*4*
Gives you a base score of...	0 points	10 points	30 points	60 points
And a bonus of...	10 points	30 points	60 points	100 points
For a total of...	10 points	40 points	90 points	160 points

TABLE 4.3. Scores Divided by 10

Landing on car number...	*1*	*2*	*3*	*4*
Gives you a base score of...	0	1	3	6
And a bonus of...	1	3	6	10
For a total of...	1	4	9	16

the adjusted base score by $n - 1$. This means that when you land, your adjusted base score is equal to

$$1+2+\cdots+(n-1)=\frac{n\,(n-1)}{2}.$$

(This isn't the first time we've seen this formula emerge! If you need a refresher on why this is true, check out the endnotes for Chapter 2.)

Also, your bonus is always equal to what your base score *would have been* if you had lasted just one more car. This means that your bonus after flying past n cars is equal to[3]

$$1+2+\cdots+n=\frac{(n+1)\,n}{2}.$$

Therefore, your total score when you land on car n is equal to[4]

$$\frac{(n+1)\,n}{2}+\frac{n\,(n-1)}{2}=n^2.$$

After a nice landing, you can take to the skies again and add to your score. This means that your final score is essentially equal to a sum of squares. It isn't exactly equal to a sum of squares because (a) the actual score is multiplied by 10, and (b) you don't land on a car in your final flight, since if you did, the game would continue. In other words, you don't get a "nice landing" bonus for your final flight.

Still, this is enough to come up with a manageable expression for your score in terms of the number of flights you take and the number of cars you pass on each flight. Suppose you play a game that involves k separate flights; on your first flight you fly past n_1 cars, on your second you fly past n_2 cars, and so on, culminating in a final flight over n_k cars. Then, based on what we've seen so far, if we let S stand for your final score divided by 10,

$$S = n_1^2 + n_2^2 + \cdots + n_{k-1}^2 + \frac{n_k(n_k - 1)}{2}.$$

That's a nice little formula. Given the duration of each of your flights (in terms of how many cars you fly over), calculating your final score is relatively straightforward.

4.3 WHAT'S IN A SCORE?

Going the other way, unfortunately, is impossible in general. That is, if you just know someone's score, there's no way to know with certainty how the player achieved that score. If someone tells you that her score (divided by 10) was 1,000,000, for instance, you'd have no idea what actions she took to reach that score. More importantly when it comes to bragging rights, you'd have no way to verify whether or not that person was telling the truth!

All is not lost, though. When you finish a game of *Sunny Day Sky*, the game provides you with a bit more information than just your raw score. It also tells you how many cars you passed in total. In other words, the game gives you S and $n_1 + n_2 + \ldots + n_k$. Let's call this second sum c (for cars).

Next question: given S and c, can you determine how the game went? In other words, can you determine the values n_i for i between 1 and k?

Next answer: no. The answer is no for a couple of reasons, some more important than others. One reason is that unless you watch the game as it happens, you'll never be able to get the ordering of the n_i for i between 1 and $k - 1$; you can permute the squares however you like and still come up with the same score. That's not such a big deal, though; we can always decide on some convention to order the n_i for recordkeeping.

TABLE 4.4. Partitions of the First Few Whole Numbers

n	p(n)	*Partitions*
1	1	{1}
2	2	{2}, {1, 1}
3	3	{3}, {2, 1}, {1, 1, 1}
4	5	{4}, {3, 1}, {2, 2}, {2, 1, 1}, {1, 1, 1, 1}
5	7	{5}, {4, 1}, {3, 2}, {3, 1, 1}, {2, 2, 1}, {2, 1, 1, 1}, {1, 1, 1, 1, 1}

After all, someone who hops over 100 cars, then 40, then 10 probably had a very similar experience as someone who hopped over 40 cars, then 100, then 10.

A more serious problem lies in the fact that we're given only two pieces of information (S and c), but we're then asking about determining k pieces of information. When $k \geq 2$, then, we're probably going to run into trouble.

Then again, there are certain restrictions on the values of n_i. Since it's impossible to *not* pass over a car, we know that each n_i is at least 1. Also, since we can't distinguish between orderings of the squares, we also may as well assume that the first $k - 1$ values are decreasing: $n_1 \geq n_2 \geq \ldots \geq n_{k-1} \geq 1$.

Unfortunately, this isn't particularly helpful for a reasonably well played game. One reason is because of the expression defining c: $n_1 + n_2 + \ldots + n_k$. Since the n_i's are all at least 1, this decomposition of c is actually a *partition* of c; that is, it's a way to write c as a sum of whole numbers.

Partitions have been studied extensively by number theorists, and quite a bit is known about them. One important question for our present purposes is *how many* partitions there are for a given number. If the number of partitions of c is small, for instance, then we can just have a computer calculate them all and check which ones could possibly lead to a score of S. So, what we're first interested in is the number of partitions that a whole number (like c) has.

This partition counting function typically goes by the name $p(n)$. You can calculate the first few values of p by hand. Table 4.4 shows how p changes for some small values of n.

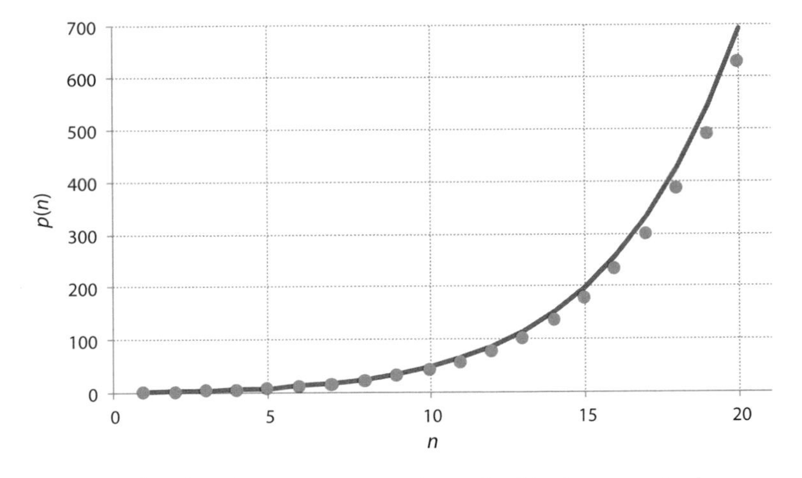

FIGURE 4.3. The blue dots represent values of $p(n)$; the curve is the approximation discovered by Hardy and Ramanujan.

YOUR SCORE
12730
TOTAL CARS
94

FIGURE 4.4. Final screen from the end of a sample *Sunny Day Sky* playthrough. Reproduced with permission from Ferry Halim.

Based on this small sample, it might seem like the number of partitions grows somewhat slowly with n. But this isn't the case; in fact, in 1918, mathematicians Hardy and Ramanujan were able to prove that as n grows, $p(n)$ grows like

$$\frac{e^{\pi\sqrt{2n/3}}}{4n\sqrt{3}}.$$

That exponential in the numerator signifies some serious growth, which you can verify by plotting this function against some values of $p(n)$ (Figure 4.3). The fast growth of $p(n)$ tells us that trying to determine n_1 up through n_k by looking at all possible partitions of c isn't feasible, at least not when c is large.[5]

Lest this get a little too abstract, let's take a look at a specific example. Suppose you play a round of *Sunny Day Sky* and wind up with a score of 12,730, while flying over 94 cars (Figure 4.4). In this case, using what you know about the scoring rules, in how many ways do you think you could score that many points while flying over that many cars?

It turns out that there are quite a few different ways: 388,920, to be precise. To be fair, this includes solutions that probably aren't that reasonable. For instance, even though it's possible for some of the n_i to equal 1, it's unlikely that any of your flights will be that short. Even if you impose greater restrictions on the n_i, though, the number of solutions is still pretty large: it's 102,536 if you assume that each n_i is at least 2, and 36,538 if you assume that each n_i is at least 3.[6]

So just knowing S and c, it turns out, doesn't tell you a whole lot. What if we throw some other info into the mix? For example, what if you keep track of the number of times you take off from the roof of a car, so that by the end of your game you know k as well as S and c? Well, depending on the value of k, this can tell you quite a bit or not very much at all.

Continuing with the current example, Figure 4.5 shows a breakdown of the number of solutions for different values of k from 1 to 53. This graph tells us, for instance, that when $k = 10$, there are more than 15,000 possible solutions (17,071, to be precise). Of these, 8,858 satisfy the additional condition that $n_i \geq 2$, and 3,876 satisfy the condition that $n_i \geq 3$.

As you can see, in most cases, knowing k doesn't help much. In particular, knowing k gives you the solution only if $k = 50$, 52, or 53.[7] Unfortunately, I'll tell you that in the round pictured in Figure 4.4, k was equal to 9. While this doesn't determine the n_i uniquely, it's better than nothing: this information reduces the number of possibilities from 388,920 to 8,902.

Let's take one more step before moving on. In general, you know S and you know c. We've seen an example of what things look like if you keep track of k. Now let's suppose that you also catch a glimpse of your final number before the "game over" screen. In other words, let's suppose you know S, c, k, and n_k.

In this case, your score, minus the contribution from your last flight, is just a sum of squares:

$$S - \frac{n_k\,(n_k - 1)}{2} = n_1^2 + n_2^2 + \cdots + n_{k-1}^2,$$

where $n_1 + n_2 + n_3 + \ldots + n_{k-1} = c - n_k$.

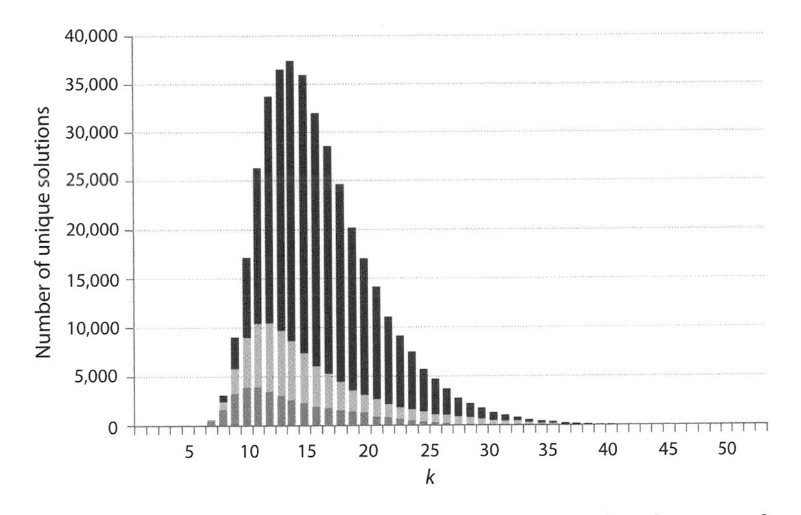

FIGURE 4.5. Number of solutions to $\{1,273 = n_1^2 + n_2^2 + \cdots + n_{k-1}^2 + \frac{n_k(n_k-1)}{2}, 94 = n_1 + n_2 + \cdots + n_k\}$ vs. k. The height of all three bars corresponds to the total number of solutions, the combined height of the orange and blue bars corresponds to solutions subject to $n_i \geq 2$, and the height of the blue bars corresponds to solutions subject to $n_i \geq 3$.

If you know S and n_k, you can calculate the left side in the above equality. So the question becomes: In how many ways can we write $S - \frac{n_k(n_k-1)}{2}$ as a sum of squares? Once again, some beautiful number theory comes into play. Just as the partition function gives us information about writing an integer as a sum of other whole numbers, there are functions that tell us the number of ways we can express a whole number as a sum of squares.

Let's ignore, for a moment, the restriction that the sum of the n_is equals $c - n_k$ and focus only on the equation involving S. When $k = 2$, things aren't very interesting:

$$S - \frac{n_2(n_2-1)}{2} = n_1^2.$$

In this case, either the left side is a perfect square and the equation has one solution, or it's not, and the equation has zero solutions.

What about the case where $k = 3$? In this case, we need to find solutions to the equation

$$S - \frac{n_3(n_3-1)}{2} = n_1^2 + n_2^2.$$

The count of integer pairs (n_1, n_2) satisfying this equality was first discovered by Carl Gustav Jacob Jacobi and is related to the numbers dividing $S - \frac{n_3(n_3-1)}{2}$. More precisely, if we let $d_1(m)$ denote the number of divisors of m that are 1 more than a multiple of 4, and $d_3(m)$ denote the number of divisors of m that are 3 more than a multiple of 4, then the number of solutions (n_1, n_2) to the equality above is equal to

$$4d_1\left(S - \frac{n_3(n_3-1)}{2}\right) - 4d_3\left(S - \frac{n_3(n_3-1)}{2}\right).$$

To make things concrete, let's return to our previous example, where S = 1,273. Suppose that you wanted to check whether or not k could equal 3, knowing that n_k = 9. In this case, this problem reduces to finding the number of ways to write 1,273 - 36 = 1,237 as a sum of two squares. Since 1,237 is a prime and is 1 more than a multiple of 4, $d_1(1, 237) = 2$ (corresponding to the divisors 1 and 1,237), $d_3(1{,}237)$ = 0. Therefore, according to the above formula, the number of ways to write 1,237 as a sum of two squares is 8:

$$9^2 + 34^2,$$
$$9^2 + (-34)^2,$$
$$(-9)^2 + 34^2,$$
$$(-9)^2 + (-34)^2,$$
$$34^2 + 9^2,$$
$$34^2 + (-9)^2,$$
$$(-34)^2 + 9^2,$$
$$(-34)^2 + (-9)^2.$$

These expressions show that there's a fair amount of double counting going on. Negative numbers contribute to the count, and pairs are counted once for every possible ordering and combination of sign. You could rightly argue that there is really only one way to write 1,237 as a sum of two squares: the square of 9 and the square of 34.[8]

Unfortunately, this doesn't provide a solution, since we also need $n_1 + n_2 + n_3 = 94$. If $n_1 = 34$ and $n_2 = n_3 = 9$, the sum of these three numbers

TABLE 4.5. Numbers $\{a, b, c, d\}$ Satisfying $a^2 + b^2 + c^2 + d^2 = 1{,}237$

{0, 0, 9, 34}	{0, 2, 12, 33}	{0, 6, 24, 25}	{0, 9, 16, 30}
{1, 4, 8, 34}	{1, 4, 14, 32}	{1, 14, 16, 28}	{2, 2, 2, 35}
{2, 3, 18, 30}	{2, 4, 16, 31}	{2, 7, 20, 28}	{2, 9, 24, 24}
{2, 14, 14, 29}	{2, 14, 19, 26}	{3, 6, 6, 34}	{4, 4, 7, 34}
{4, 4, 23, 26}	{4, 8, 14, 31}	{4, 14, 20, 25}	{4, 16, 17, 26}
{6, 6, 18, 29}	{6, 7, 24, 24}	{6, 15, 20, 24}	{7, 8, 10, 32}
{7, 12, 12, 30}	{7, 16, 16, 26}	{8, 8, 22, 25}	{8, 10, 17, 28}
{8, 17, 20, 22}	{9, 16, 18, 24}	{10, 10, 14, 29}	{10, 10, 19, 26}
{10, 13, 22, 22}	{10, 16, 16, 25}	{12, 12, 18, 25}	{13, 14, 14, 26}
{14, 14, 19, 22}	{14, 16, 16, 23}		

is only 52, which is too small. So this doesn't provide a solution to our example from before, and we can conclude that if $c = 94$, $S = 1{,}237$, and $n_k = 9$, k can't equal 3.

We can take a similar approach with other values of k. For example, when $k = 5$, we need to find the number of ways to write $S - \frac{n_5(n_5-1)}{2}$ as a sum of four squares. This question was also explored by Jacobi, who showed that the number of ways to write an integer m as a sum of four squares is equal to 8 times the sum of divisors of m which are not also divisible by 4. For example, if we want to write 1,237 as a sum of four squares, there are 9,896 ways to do this. This is because neither divisor of 1,237 is divisible by 4, so the sum of these divisors is $1{,}237 + 1 = 1{,}238$, and $1{,}238 \times 8 = 9{,}896$.

Of course, as before, this tally also includes substantial double counting. Ignoring negative numbers and different orderings, there are really only 38 unique ways to write 1,237 as a sum of four squares (listed in Table 4.5). Unfortunately, as before, none of these potential solutions correspond to a partition of 94.

What about different values of k? Formulas for counting the number of ways to write a whole number as a sum of m squares are known for other values of m, but unfortunately the formulas become increasingly complicated as m grows. So while this approach will sometimes give us only a small number of options to choose from, it's also only feasible for a restricted number of values of k and requires that we know n_k.[9] No matter how you slice it, figuring out the details of someone's playthrough from just a few pieces of information isn't feasible. But in the search lies some beautiful mathematics—not a bad consolation prize, all things considered.[10]

4.4 *THREES!* COMPANY

Mobile gaming has many other examples of interesting scoring mechanics; we could fill an entire book with explorations of how some of these games are scored. To keep things moving, we'll just look at a couple more examples.

In early 2014, indie game development shop Sirvo (consisting of just three people: Asher Vollmer, Greg Wohlwend, and Jimmy Hinson) released a beautifully designed game called *Threes!*, which follows the trajectory of many classic games by being simple to learn but incredibly difficult to master. It quickly inspired a number of rip-offs,[11] but its beautiful design set it apart, and Apple named it the best iPhone game of 2014.

The game presents players with a 4×4 grid along with some numbered cards. Players can swipe left, right, up, or down; a swipe in one of these directions will cause all cards to move one spot over in that direction (unless they are unable to move, in which case they will remain in their current position). After every swipe, a new card appears on the board.

With a new card appearing after every move, and only 16 spaces on the board, the game would end rather quickly were it not for its central mechanic: you can combine cards together to form new ones with higher numerical values. "1" cards combine only with "2" cards (and vice versa); combine a pair and you'll get a "3." After that, every card must combine with another of its own type: two "3" cards combine to form a "6," two "6" cards combine to form a "12," and so on. The sequence of boards in Figure 4.6 illustrates this mechanic.

The game proceeds in this way until the board is filled and the player has no more moves. The player then receives a score based on the cards on the board. The "1" and "2" cards are worth 0 points, "3" cards are worth 3, "6" cards are worth 9, "12" cards are worth 27, and in general, a card numbered 3×2^n is worth 3^{n+1} points.

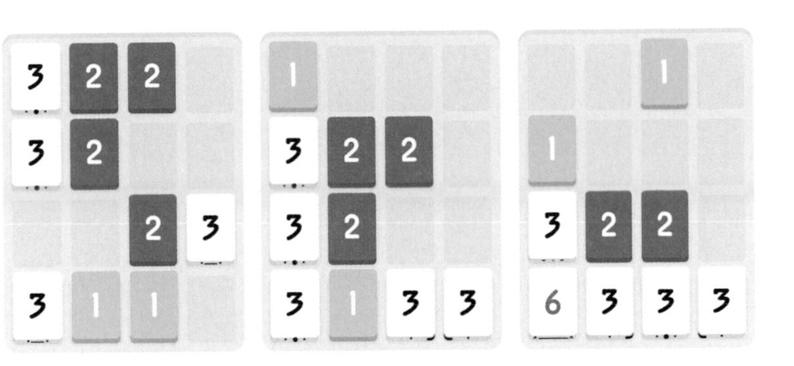

FIGURE 4.6. On the left is a starting board. One downward swipe yields the middle board (note that the "2" in the third row of the starting board combines with the "1" directly below it to form a "3"). Another swipe down yields the rightmost board. Screenshots from *Threes!*, reproduced with permission from Asher Vollmer.

Unlike *Sunny Day Sky*, scores in *Threes!* are shared on worldwide leaderboards, so you can easily see how you compare to other players. When you view the rankings, though, you see only the score, not the final board. So, as with *Sunny Day Sky*, we can ask: given the final score, what can we say about the board that led to it?

Of course, just as we'll never be able to determine the ordering of the k jumps in *Sunny Day Sky*, we'll never be able to determine the specific positioning of the cards in a *Threes!* game just from the final score. For example, even if a player has terrible luck and finishes with a score of 3 points, there are 32 possible boards that would lead to such a score, since there are 16 places where the "3" card could be, and the other cards could all be "1"s or "2"s (Figure 4.7).

Rather than trying to determine the layout of the board, then, let's focus on a slightly simpler problem: given the score, can we determine *how many* cards of each point value must have been on the board? For example, if the final score is 3, we know that the board must have one "3" card and 15 cards worth 0 points—in other words, ignoring some details of the board layout, there's essentially only one way to score 3 points. Similarly, there's only one way to score 6 points: you must have two "3" cards and 14 cards worth nothing. However, there are two ways to score 9 points: your board can have three "3" cards and 13 cards worth nothing, or it can have one "6" card and 15 cards worth nothing.

FIGURE 4.7. Two examples of boards that would lead to a score of 3. It should be noted that while these are theoretically possible, in practice they're impossible given the way the board is initialized (you always start with some "3" cards, some "2" cards, and some "1" cards). Screenshots from *Threes!*, reproduced with permission from Asher Vollmer.

With *Sunny Day Sky*, we saw that knowing the score and the number of cars you hopped was actually woefully insufficient if you wanted to know more about the playthrough. But what about *Threes*? In order to answer this question, we'll need a more systematic approach.

4.5 A MATHEMATICAL MODEL OF *THREES!*

There are twelve cards in *Threes!* that earn you points: the "3" card, the "6" card, and so on, all the way up to the "6144" card. If we let c_1 denote the number of "3" cards, c_2 denote the number of "6" cards, and so on, then the player's final score, S, is given by

$$S = 3c_1 + 3^2c_2 + 3^3c_3 + \ldots + 3^{12}c_{12}.$$

Moreover, we know a few things about the c_i. For one, the sum of them must be less than 16, since there are only sixteen spots on the board. In other words,

$$c_1 + c_2 + c_3 + \ldots + c_{12} \leq 16.$$

There's a second condition that almost all of the c_i must obey. Notice that no final board can have more than eight cards with the same (nonzero) point value; if a board has nine or more cards of the same

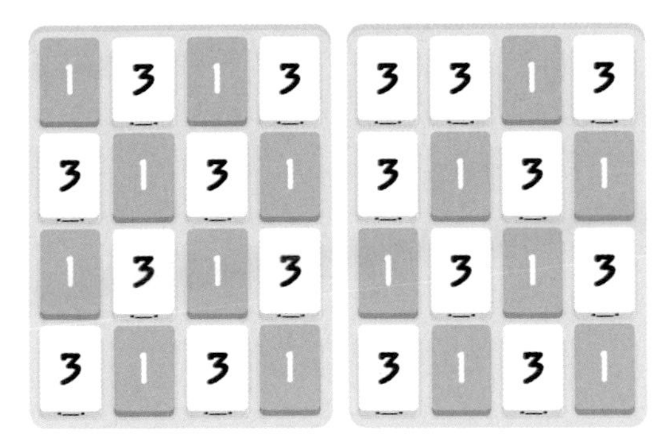

FIGURE 4.8. Two boards to highlight the second restriction on each c_i. On the left, there are eight "3" cards, and no moves for the player, so the game is over; on the right, there are nine "3" cards, but the player can swipe in any direction to keep the game going. Screenshots from *Threes!*, reproduced with permission from Asher Vollmer.

point value, the player could swipe the board in some direction in order to combine cards and continue playing.

In other words, almost all of the c_i must be between 0 and 8. The only exception is c_{12}; since there is no card higher than the 6144 card, if you ever get two 6144 cards together, you won't be able to combine them. As developer Asher Vollmer told me in an e-mail, "theoretically you could have sixteen 6144s on the board at the end of the game, but it is practically impossible."

Therefore, our original question now becomes one of determining, for a given positive score S, how many whole number values of $c_1, c_2, \ldots, c_{12}$ there are such that

$$S = 3c_1 + 3^2c_2 + 3^3c_3 + \ldots + 3^{12}c_{12},$$

$$c_1 + c_2 + c_3 + \ldots + c_{12} \leq 16,$$

$$0 \leq c_1, c_2, c_3, \ldots, c_{11} \leq 8.$$

$$0 \leq c_{12} \leq 16.$$

Let's let $n(S)$ stand for the number of solutions to this set of equations and inequalities for a positive score S. In the examples above we saw that $n(3) = 1$, $n(6) = 1$, and $n(9) = 2$. In fact, when S is fairly small, Table 4.6 shows that the pattern is relatively predictable.

TABLE 4.6. Values of $n(S)$ When S Is Small

S	$n(S)$
3	1
6	1
9	2
12	2
15	2
18	3
21	3
24	3
27	4
30	4
33	4

Unfortunately, this pattern can't continue forever. For one thing, S has a maximum: when $c_{12} = 16$, we get $S = 8{,}503{,}056$, and since this is the only way for S to be so large, we have $n(8{,}503{,}056) = 1$. Furthermore, it's not the case that every multiple of 3 between 8,503,056 is a possible value for S; for example, $n(39{,}363) = 0$, since the minimum number of cards you'd need to achieve a score of 39,363 is 17:

$$39{,}363 = \mathbf{2} \times 3 + \mathbf{2} \times 3^2 + \mathbf{2} \times 3^3$$
$$+ \mathbf{2} \times 3^4 + \mathbf{2} \times 3^5 + \mathbf{2} \times 3^6$$
$$+ \mathbf{2} \times 3^7 + \mathbf{2} \times 3^8 + \mathbf{1} \times 3^9.$$

Suffice it to say, the function $n(S)$ isn't easy to write down by hand. But, since it has only a finite domain, it's possible to have a computer do the grunt work for us. Figure 4.9 shows a graph of $n(S)$, where S runs over all multiples of three from 3 to 8,503,056.

While a description of $n(S)$ is hard to write down as a function rule, the general trend shouldn't be so surprising: on the whole, $n(S)$ increases for a time, then gradually tapers off. The largest value of $n(S)$ is 393, when $S = 730{,}728$. In other words, if that's your final score from a game of *Threes!*, there are 393 ways you could have achieved that score. Given that this is the maximum value of $n(S)$, it seems clear that

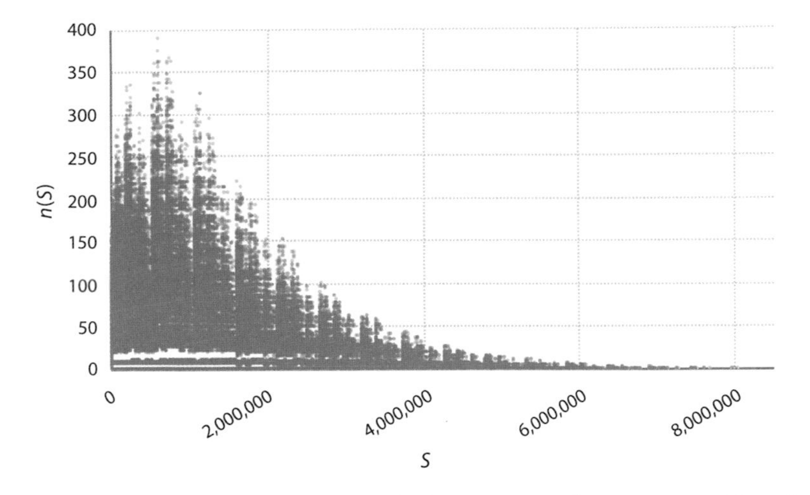

FIGURE 4.9. A graph of $n(S)$ vs. S. Each dot corresponds to one possible value of S.

knowing the final score in *Threes!* gets you closer to finding a unique answer than it does in *Sunny Day Sky*.

In fact, of the $8{,}503{,}056 \div 3 = 2{,}834{,}352$ possible values of S in the graph above, 1,769,794 of them have $n(S) = 0$. In other words, there are only 1,064,558 possible final scores from a game of *Threes!* Among these, 342,359 of them have $n(S) = 1$, meaning that in roughly 32.2% of the possible cases, knowing S is enough to uniquely determine the values $c_1, c_2, \ldots, c_{12}$. Not too shabby.

Of course, not every score is equally likely, so this analysis doesn't give a completely accurate picture. Indeed, most of those unique scores come at the higher end of the spectrum, and it's unlikely that anyone's score will ever reach those heights. In fact, according to the developers of *Threes!*, more than 75% of players haven't seen the four highest scoring cards (Figure 4.10).

Because of this, for the typical player it may be better to restrict the above analysis to only a subset of all the available cards. For instance, suppose we're interested only in scores involving cards up to 384. Our question is still the same: given a score S, in how many ways is it possible to come up with that score? Let's call our function $m(S)$ to distinguish it from $n(S)$; m counts solutions using the first eight cards, n counts solutions using all twelve.

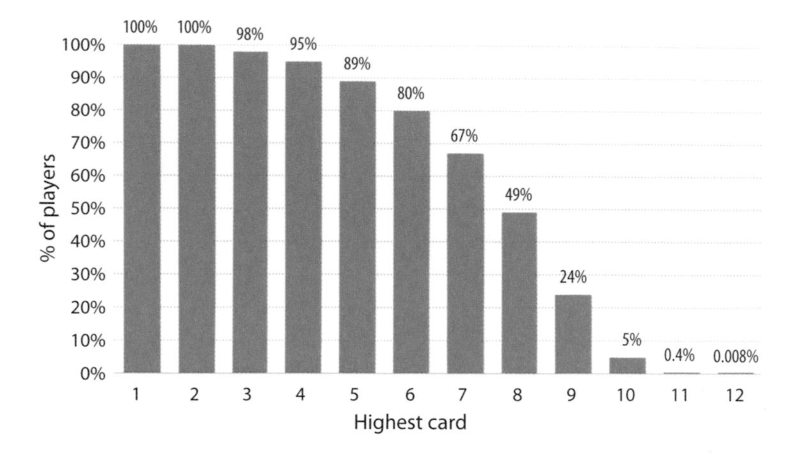

FIGURE 4.10. A graph of the highest scoring card that each player has seen. For instance, 49% of players (as of February 2015) have seen the 384 card, but only 24% have seen the 768 card. Data from *Threes!*, with permission from Asher Vollmer.

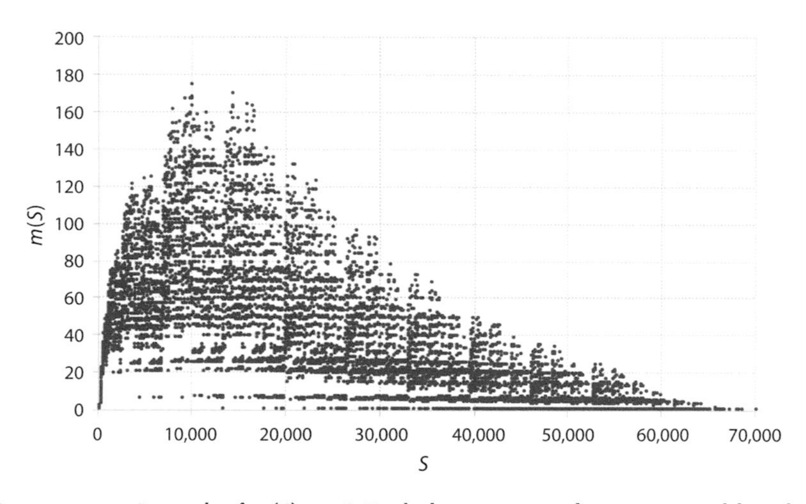

FIGURE 4.11. A graph of $m(S)$ vs. S. Each dot corresponds to one possible value of S. Here the most common score is 9,828, with $m(9{,}828) = 175$.

Using only the first eight cards, the player's maximum score is $8 \times 3^7 + 8 \times 3^8 = 69{,}984$, and since S must be divisible by 3, this gives us $69{,}984 \div 3 = 23{,}328$ possible values for S. Figure 4.11 shows what $m(S)$ looks like over that range of values. The general shape of $m(S)$ looks similar to the shape for $n(S)$. However, when you dig into the numbers, it's clear

TABLE 4.7. Comparison of $m(S)$ and $n(S)$

Function	*# of times = 1*	*% of times = 1*	*Average value*
$n(S)$	342,359	32.2%	10.5
$m(S)$	2,651	14.7%	29.3

that having fewer cards makes pinning down the values of c_i just from S more difficult.

As with our earlier counting function, $m(S)$ is occasionally zero. More precisely, $m(S) = 0$ for 4,042 values of S within our range, so there are actually $23{,}328 - 4{,}042 = 19{,}286$ possible values of S. Out of these, $m(S) = 1$ a total of 2,651 times. In other words, the values of c_i can be determined just from S roughly 14.7% of the time.

Table 4.7 highlights some of the comparisons between the functions m and n. If we treat all final boards equally, it's much more common to be able to determine the cards on the board if all twelve cards are in play rather than just eight. Also, the average value of n is almost one-third what it is for m. In other words, when we restrict to eight cards, for each value of S there are, on average, more than twenty-nine different sets of cards that yield S, versus an average of 10.5 sets when you allow for all twelve cards.

In summary, restricting to a smaller number of cards makes determining the c_is just from the final score more difficult. In either case, though, it's much easier to infer information from the score in this game than it is in *Sunny Day Sky*.[12]

4.6 INVALID SCORES

We've explored a lot of math, but it may seem like a significant investment for very little payoff. What's one to do, for example, with all this esoteric knowledge about the information hidden in the final scores of games like *Sunny Day Sky* and *Threes!*?

Well, the mathematics itself should feel rewarding in its own right (I hope!). But for games like *Threes!*, which have global online leaderboards, there's also a practical application to understanding how the scoring works. It allows us to assess whether or not anything fishy is going on with the online rankings.

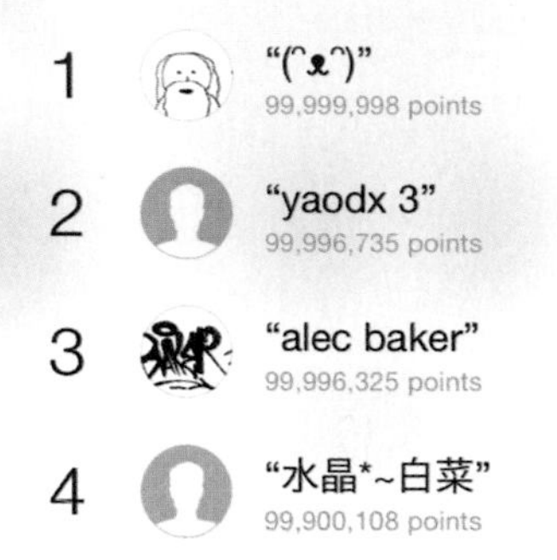

FIGURE 4.12. The posted high scores for *Threes!* a few weeks after its release. Screenshot reprinted with permission from Apple Inc.

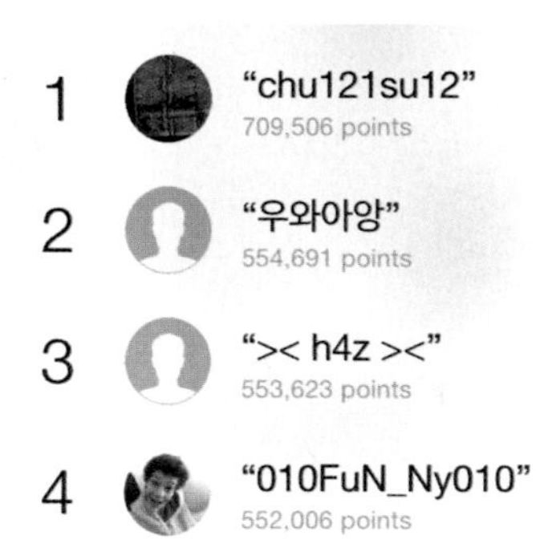

FIGURE 4.13. High scores roughly a week later. Screenshot reprinted with permission from Apple Inc.

Some people play these games hoping to get a score that will propel them to the top of the leaderboards. But not all high scores are obtained by playing the game fair and square. For example, shortly after *Threes!* was released, the recorded high scores (shown in Figure 4.12) were highly suspect.

The scores should arouse your suspicions for a couple of reasons. The first is that they're all are way too high; based on our analysis, they're much larger than any player could ever obtain. The second (and arguably more egregious) reason is that not all of the scores are divisible by 3. Clearly, something fishy is going on.

It didn't take long for these phony scores to be scrubbed from the record books. About a week later, the top four scores were much more

TABLE 4.8. Analysis of the Scores in Figure 4.13

S	$n(S)$
709,506	230
554,691	79
553,623	78
552,006	182

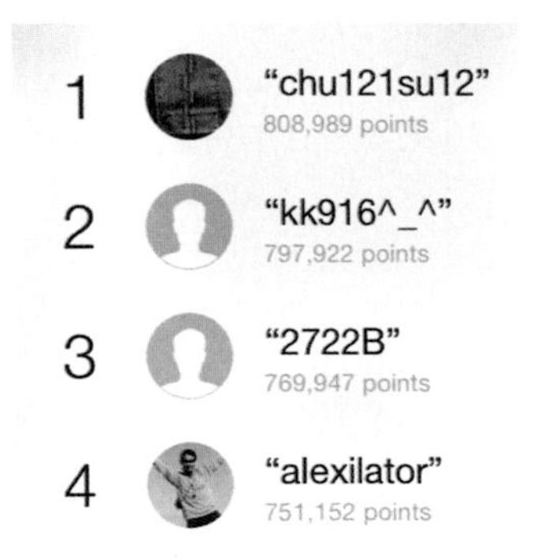

FIGURE 4.14. High scores a few months later. Screenshot reprinted with permission from Apple Inc.

TABLE 4.9. Analysis of the Scores in Figure 4.14

S	$n(S)$
808,989	9
797,922	12
769,947	86
751,152	133

reasonable (Figure 4.13). Unlike before, each of these scores is possible. In fact, Table 4.8 shows $n(S)$ for each value of S in Figure 4.13.

A few months later, the top four scores had increased even further (Figure 4.14). Moreover, the values of $n(S)$ had become lower (Table 4.9).

Of course, it isn't always true that when S increases, $n(S)$ decreases. Nor is it true that once one set of suspicious scores is scrubbed, another set won't appear in its place. When I checked back a few months later, the top four scores looked suspicious once again (Figure 4.15).

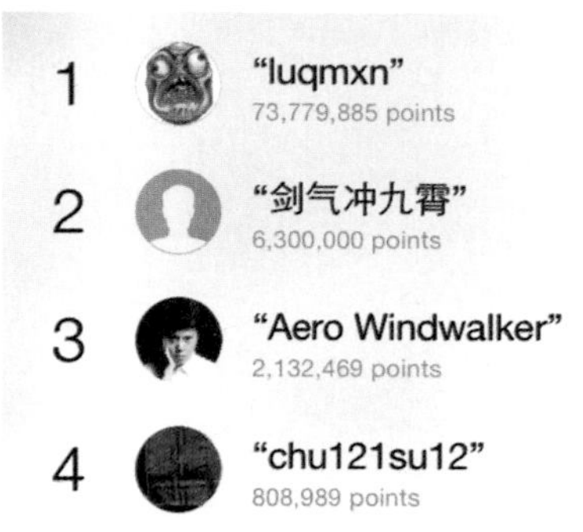

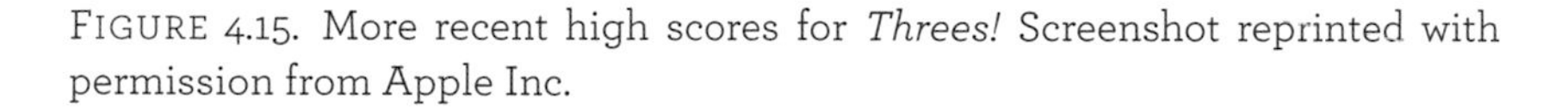

FIGURE 4.15. More recent high scores for *Threes!* Screenshot reprinted with permission from Apple Inc.

The scores in Figure 4.15 are all interesting, but for different reasons. The first is clearly a fake; it's nearly ten times larger than the maximum possible score. The second is within the range of possible values, but it turns out to be a fake too. In fact, in order to get a score of 6,300,000, you'd need a minimum of twenty-four cards:

$$6{,}300{,}000 = \mathbf{1} \times 3^2 + \mathbf{2} \times 3^3 + \mathbf{2} \times 3^4 + \mathbf{2} \times 3^5 + \mathbf{1} \times 3^6 + \mathbf{2} \times 3^9 + \mathbf{1} \times 3^{10} + \mathbf{2} \times 3^{11} + \mathbf{11} \times 3^{12}.$$

So while the second score isn't as obviously a forgery as the first, it's a forgery nevertheless.

Let's skip the third score for a moment and spend some time at the fourth score: 808,989. As mentioned above, there are nine ways a player could achieve this score without cheating. The developers of *Threes!* have also gone on record as saying that they believe the score is legitimate.[13] Assuming that it is, can we say anything about which of the nine possible collections of cards is the most likely?

To explore this question, we need to know what the nine possible cases look like. The cases are enumerated in Table 4.10.

Here are couple of things to notice: Case 1 has the fewest number of point-scoring cards (13, vs. 15 in all the other cases).[14] Also, it is the only case in which c_i is no larger than 2 for each i. Since it's more difficult to have large values of c_i in the final board configuration, one might reasonably infer that case 1 is the most likely way in which the player

TABLE 4.10. Possible Ways to Score 808,989 Points

Case	c_1	c_2	c_3	c_4	c_5	c_6	c_7	c_8	c_9	c_{10}	c_{11}	c_{12}
1	2	1	1	0	2	2	0	0	2	1	1	1
2	5	0	1	0	2	2	0	0	2	1	1	1
3	2	4	0	0	2	2	0	0	2	1	1	1
4	2	1	1	3	1	2	0	0	2	1	1	1
5	2	1	1	0	5	1	0	0	2	1	1	1
6	2	1	1	0	2	2	0	3	1	1	1	1
7	2	1	1	0	2	2	0	0	5	0	1	1
8	2	1	1	0	2	2	0	0	2	4	0	1
9	2	1	1	0	2	2	0	0	2	1	4	0

obtained a final score of 808,989, while cases like 5 and 7 (in which values of c_i are as high as 5) are less likely.

We've seen one clearly fraudulent score, one subtly fraudulent score, and one score that seems reasonable. That leaves us with the third highest score: 2,132,469. Even though it's significantly higher than the fourth-place score, there are fifty-two ways to achieve it, compared to only nine ways for 808,989. Among these fifty-two possibilities, the minimum number of point-scoring cards you need is nine:

$$2,132,469 = \mathbf{1} \times 3^2 + \mathbf{2} \times 3^3 + \mathbf{1} \times 3^4 + \mathbf{1} \times 3^8 + \mathbf{4} \times 3^{12}.$$

Does a final board with four 6,144 cards seem realistic, when the next-highest score has only one? Many of the fifty-two options require four 6,144 cards; the minimum number you can get away with is 2, but in such a case you'd need at least five 3,072 cards, which doesn't seem very reasonable.

If you doubt the validity of this score, you're not alone. This score has long since been removed.

4.7 LOWEST OF THE LOW

So far, we've been focusing on high scores. But, like golf, there are games in which a high score is a *bad* thing. Instead, what you really might be aiming for is a low score, or, in the case of timed events, a low time.[15]

In fact, in recent years, *speedrunning*—competitively trying to complete a game in the fastest possible time—has exploded in popularity. Live streaming sites like Twitch.tv have made video games more of a spectator sport than ever before, and dedicated players devote countless hours to shaving fractions of a second off of their best times.[16]

Aside from speedrunning, there are also examples of game mechanics that encourage a low score rather than a high score. One example comes from the game series *Fallout*. If you've ever wished to experience a 1950s aesthetic combined with a postapocalyptic landscape filled with death and decay, then *Fallout* may be right up your alley. What began with the well-received release of *Fallout* in 1997 has since evolved into one of the most popular video game franchises around. The latest entry, *Fallout 4*, was released in 2015 to critical and commercial acclaim.

Fallout 4, like many of its predecessors, features many side quests and minigames meant to hold the player's attention long after the main campaign has ended. One is a game in which players attempt to "hack" computers in order to find loot or unlock secrets near their location.

The hacking minigame was first introduced in *Fallout 3*. The idea is simple: the player is presented with a list of possible passwords to hack the computer. For example, you might be given a list of twelve five-letter words: one of those words is the password, and the rest are duds. Next, you're given four chances to guess the correct password. Every time you guess incorrectly, though, the computer tells you how many letters in your guess match a letter in the same position in the actual password; this is referred to as your guess's *likeness*. So, for example, if the password is "class" and you guess "atlas," the computer would tell you that the likeness is 1 (only the final "s" counts, since that's the only matching letter that is also in the right position).

In other words, the goal here is to find the password in as few guesses as possible. If you can't guess the password in four tries, the computer locks up for good. Rather than trying to *maximize* some high score then, in this case players must *minimize* the number of guesses they need to unlock the computer.

Let's look at a more specific example. One day, as I was playing *Fallout 4*, I came across a locked computer that gave me the following list of words as potential passwords:

piled, viral, sound, fence, fists, forth, grass, start, shape, sewer, agent, parts.

Suppose I started by guessing *viral*, and the game told me that the likeness was 0. Based on this, I could safely eliminate *piled*, (matching "i"), *forth*, (matching "r"), and *parts* (matching "r"). Effectively I've reduced the word list from twelve to eight. Not bad.

Play this game long enough, and you'll likely develop some strategies to help you win. It's difficult, however, to implement more sophisticated strategies by hand. The game itself has no way to mark a word as a dud, so you have to use a fair amount of mental bandwidth keeping track of which words are viable solutions and which aren't.

So, let's do what any good mathematician would do and try to come up with a more systematic approach. To begin, we'll take a look at a simplified example. Suppose you're given a list of five four-letter words: *hard, herd, hero, zero,* and *jump*. What approach should you use to find the correct password?

Below are a few options. For each of these strategies, once you make a guess and receive your guess's likeness score, the idea is that you should remove any words that can't be the password and repeat the strategy.

1. Guess randomly. This strategy is probably the most straightforward. With this approach, each of the five words has an equally likely chance of being your guess.
2. Guess the word that is "closest" to the others. In some sense, which we're about to make precise, "herd" is closer to "hard" than "hard" is to "jump." If you have a notion of distance on these words, you could use it to determine the word that is "closest" to all the rest. This word, as it would be likely to be close to the password, might make for a good candidate.
3. Guess the word that is "farthest" from the others. In the event that the password is also far, maybe this would be a better approach.
4. Guess the word that, on average, will allow you to eliminate the most words from the list.

TABLE 4.11. Distance between Words

	hard	*herd*	*hero*	*zero*	*jump*
hard	0	1	2	3	4
herd	1	0	1	2	4
hero	2	1	0	1	4
zero	3	2	1	0	4
jump	4	4	4	4	0
Average	2.0	1.6	1.6	2.0	3.2

The differences between these strategies (especially the second and fourth) may not be immediately clear. So let's consider each one within the context of our current example.

First, we'll need to define a notion of distance on two words. Assuming that the words have the same length, we'll use the Hamming distance, which is just a count of the number of letters that the two words do not have in common (including position). For example, *boss* and *moss* have a distance of 1, since they differ only in the first character. *Trap* and *harp* have a distance of 3; they have many letters in common, but only the *p* is a common letter in the same position in both words. Two words have a distance of 0 precisely when they are the same word.[17] If you'd rather think in terms of likeness, the Hamming distance between two words is just the difference between their common length and their likeness.

Returning to our four strategies, let's make a table of our words and calculate the distances between them. The data are summarized in Table 4.11.

Based on the averages in Table 4.11, the words that are closest to every other word in the set, according to the Hamming distance, are "herd" and "hero." Similarly, the farthest word is "jump." As for the word with the highest number of expected eliminations, for that we'll need to do a bit more calculation.

To begin, let's suppose that each word has the same probability of being the password. If you guess "jump," there's a 20% probability that you're correct. In this case, you'll be able to eliminate every word from the list of potential guesses, since you'll have already guessed the password.

TABLE 4.12. Expected Number of Eliminations for Each Word

hard	*herd*	*hero*	*zero*	*jump*
4.2	3.8	3.8	4.2	1.8

However, there's an 80% chance that "jump" isn't the password. In this case, you'll be able to eliminate only "jump" from the list of words, since your likeness score is guaranteed to be 0, and every word besides "jump" shares at least one character with every other word.

Putting these facts together means that the expected number of eliminations from choosing "jump" is equal to $0.2 \times 5 + 0.8 \times 1 = 1.8$.

On the other hand, suppose you choose "hard" as your word. In this case, there's a 20% probability you'll be able to remove five words from the list, and an 80% chance you'll be able to remove four words from the list. That's because if "hard" isn't the word, its likeness will immediately determine which word is the password. Put another way, the other four words each have a unique distance to "hard."

In this case, then, the expected number of eliminations from choosing "hard" is equal to $0.2 \times 5 + 0.8 \times 4 = 4.2$.

In fact, "hard" and "zero" have the highest number of expected eliminations. Table 4.12 has the data on the other words.

Basically, if you go with the closest word strategy, you should choose "herd" or "hero." If you go with the farthest word strategy, you should choose "jump." And if maximizing the expected number of eliminations is more your cup of tea, you should choose "hard" or "zero." Of course, guessing randomly could give you any one of these answers.

But which strategy is the best for a wide variety of word lists? There are a couple of different metrics we could use to determine the "best" strategy. One way would be to calculate what percentage of the time each strategy is able to identify the password in four guesses or fewer. Alternatively, you could ask what the expected number of guesses is for each strategy. Let's do both!

In order to test these strategies, we also need word lists. Let's start by using a few lists from *Fallout 4*, including the one we saw earlier:

List 1: *piled, viral, sound, fence, fists, forth, grass, start, shape, sewer, agent, parts.*

TABLE 4.13. Statistics for Different Strategies and Different Word Lists

	Strategy			
Words	*Random*	*Closest*	*Farthest*	*Max. Elim.*
List 1	3.06 Guesses (92.79%)	2.92 Guesses (91.67%)	3.50 Guesses (75.00%)	2.75 Guesses (100%)
List 2	2.83 Guesses (96.21%)	2.58 Guesses (100%)	3.25 Guesses (83.33%)	2.58 Guesses (100%)
List 3	2.78 Guesses (97.93%)	2.58 Guesses (100%)	3.08 Guesses (91.67%)	2.58 Guesses (100%)

List 2: *energy, mutter, warned, atrium, second, carved, forced, hungry, depend, heated, mirror, stream.*

List 3: *varying, cistern, expects, attends, bottles, torches, limited, corners, fortify, despite, session, durable.*

Table 4.13 shows some data on how each strategy performs with respect to each of these lists. For the random guessing strategy, I've simulated 100,000 trials of the game for each possible password, or 1.2 million trials in all. The other strategies are completely deterministic, so you need to play the game only once for each password to know how many guesses you'd need to make.

Based on the data in Table 4.13, it's clear that selecting the farthest word is your worst option. Guessing randomly does surprisingly well in terms of win percentage, though the average number of guesses you'll need is higher than with the closest word or maximum expected elimination strategies. And the only strategy to guarantee a win 100% of the time is maximum expected elimination. So is this the one strategy to rule them all?

Before we declare a winner, it's probably worth looking at some other sets of words. After all, the word lists in *Fallout 4* are probably deliberately chosen so that there's a relatively high degree of likeness between words in the list. If every word had a likeness of 0 compared to every other word, all you could do to try and guess the password is blindly guess. Put another way, if all the likeness scores are 0, the likeness score effectively tells you nothing.

TABLE 4.14. Different Strategies Compared across Randomly Generated Word Lists

	Strategy			
Words	*Random*	*Closest*	*Farthest*	*Max. Elim.*
List 1	3.19 Guesses (85.92%)	2.88 Guesses (96.04%)	3.82 Guesses (64.19%)	2.83 Guesses (97.50%)
List 2	3.06 Guesses (90.64%)	2.80 Guesses (97.42%)	3.60 Guesses (71.31%)	2.73 Guesses (98.94%)
List 3	2.96 Guesses (93.67%)	2.73 Guesses (98.18%)	3.43 Guesses (77.77%)	2.65 Guesses (99.57%)

So let's look at some lists of randomly generated words. Online, it's not too hard to find lists of words of a certain length; using those lists, I had my computer quickly whip up 10,000 word lists of twelve words each. I did this three times: once for five-letter words, again for six-letter words, and finally for seven-letter words. The results are in Table 4.14 (this time I ran only 1,000 trials for the random strategy, not 100,000, since I had to do 1,000 trials across 30,000 different word lists).

In this case, the farthest word strategy is the poorest performer by far. This is true both in terms of the average number of guesses needed to get the right answer, and in terms of the win percentages. In contrast, the maximum expected eliminations strategy is still the best, but simply selecting the closest word in the word list is fairly comparable. Both are better than guessing randomly. And no matter the strategy, metrics improve as the number of letters per word increases.

Of course, if one wanted to, there's still plenty more to dig into here. Can you think of any other strategies? What if you allow for word lists that are longer than twelve? What if you look at words of lengths less than five or greater than seven? What if you allow for any sequence of letters, not just English words? What if you allow for word lists that contain words of varying lengths? How do you calculate distances in this case?[18]

And lest you think that all of this work has no application outside the realm of *Fallout*, the idea of measuring distance between two pieces of text extends far beyond this one example. Indeed, with slightly more sophisticated notions of distance than we've discussed here, there are applications in biology and computer science, too. Any time you've used a spellchecker or Google has corrected a typo for you, you can bet that under the hood there were some distance calculations being performed on pieces of text.[19]

In practice, writing an algorithm to find the most efficient way to guess a password in *Fallout* may not be worth the effort. But for a particularly challenging list of words, a little extra forethought can make all the difference.

4.8 HIGHEST OF THE HIGH

We've talked about the mechanics of high scores for different games, and how sometimes an understanding of those mechanics can help us figure out whether or not a posted high score is legitimate. We've also examined a scenario in which a *low* score is the thing to shoot for rather than a *high* score.

Here's one final question: Practically speaking, what's the highest high score one can achieve in a particular game? In *Threes!*, for example, we saw that the highest theoretical score was 8,503,056. However, it's effectively impossible to come anywhere near this number. So if you're aiming for the top score—in *Threes!* or in any other game—what's a reasonable goal to set for yourself?

In some cases, it's probably not worth trying to break the record. For example, there's a maximum possible high score in the arcade version of *Pac-Man*, too: 3,333,360 points. But unlike in *Threes!*, it's possible to achieve this high score if you're devoted enough. And indeed, since the first "perfect" *Pac-Man* game was completed by arcade aficionado Billy Mitchell in 1999, a handful of others have joined his ranks. So if you want to break the high-score record in *Pac-Man*, you're out of luck. At best, you could match the current record, but do so in less time. Speedrunning strikes again!

In other words, when it comes to aspirations of getting your name on the top of the leaderboard, the first question might not be, What's a reasonable goal to set for myself? Instead, it may be, Is this game even worth tackling at all?

In fact, questions like these have been addressed in the context of sports. Researchers have predicted limits to the fastest time we will ever see in the men's 100-m sprint, for example. They've also examined which world records in Olympic events are closest to their theoretical limits, and which have more room for improvement.[20]

Unfortunately, the amount of data on high scores for a particular video game pales in comparison to the amount of data on, say, records for the men's 100-m sprint. And the high-score data that do exist, as we've already seen, aren't always the most trustworthy. This lack of data can make future prediction more difficult. And while there are organizations that track records with an eye toward verifying high scores, this upgrade in quality comes at the cost of quantity: by and large, high-score databases for an individual game are relatively small, making analysis difficult.

In some cases, though, there's enough information to at least make some simple predictions. One example is the classic arcade game *Donkey Kong*. Originally released by Nintendo in 1981, *Donkey Kong* first introduced the world to the eponymous giant gorilla. It was also the first game to feature Mario, though at the time he simply went by the far-too-literal name "Jumpman."

In the game, Jumpman chases the nefarious Donkey Kong higher and higher up a construction site in order to rescue his girlfriend, Pauline. *Donkey Kong* was, and still is, one of the most enjoyable arcade games of all time. No wonder, then, that there's still a relatively strong fan base for a game that's more than thirty-five years old.

The game is so popular, in fact, that it was the subject of a 2007 documentary *The King of Kong: A Fistful of Quarters*. The movie highlighted a worldwide community of fans who still play classic arcade games and compete with one another for the highest score. A number of games are showcased, but, as the title suggests, *Donkey Kong* is the film's primary focus. This is due in large part to a rivalry between two men—Steve Wiebe and the aforementioned Billy Mitchell—who competed fiercely for the *Donkey Kong* high score for a few years in the mid-2000s.

Mitchell owned the high score by a comfortable margin from 1982 through 2000. However, in recent years the competition for his crown has grown fierce, and Mitchell hasn't owned the high score since 2010. Table 4.15 shows how the record has changed over time since Mitchell's 1982 record of 874,300 points was broken.

TABLE 4.15. *Donkey Kong* High Scores over Time

Date	*High Score*	*Player*[21]
8/17/2000	879,200	Tim Sczerby
5/7/2004	933,900	Billy Mitchell
6/3/2005	985,600	Steve Wiebe
6/4/2005	1,047,200	Billy Mitchell
8/3/2006	1,049,100	Steve Wiebe
7/13/2007	1,050,200	Billy Mitchell
2/26/2010	1,061,700	Hank Chien
7/31/2010	1,062,800	Billy Mitchell
8/30/2010	1,064,500	Steve Wiebe
12/27/2010	1,068,000	Hank Chien
2/27/2011	1,090,400	Hank Chien
5/18/2012	1,110,100	Hank Chien
7/25/2012	1,127,700	Hank Chien
11/1/2012	1,138,600	Hank Chien
9/5/2014	1,141,800	Robbie Lakeman
12/1/2014	1,144,800	Robbie Lakeman
6/24/2015	1,158,400	Robbie Lakeman
9/17/2015	1,170,500	Wes Copeland
9/18/2015	1,172,100	Robbie Lakeman
10/21/2015	1,177,200	Robbie Lakeman
1/4/2016	1,190,000	Wes Copeland
4/11/2016	1,190,200	Robbie Lakeman
4/19/2016	1,195,100	Wes Copeland
5/5/2016	1,218,000	Wes Copeland

Mitchell and Wiebe duked it out over the top score over a period of several years, though neither one has held the top spot recently. The more recent defenders of the crown have been Robbie Lakeman and Wes Copeland. In fact, when Copeland broke the 1.2 million point threshold in 2016, some news outlets reported that he had achieved a "perfect" score.[22] In principle, however, there's still room for improvement. But how much room, exactly, and is it worth trying to usurp Copeland's throne?

To answer this question, we need to fit a model to the data. One approach is to assume that the high score data are roughly *logistic*, or

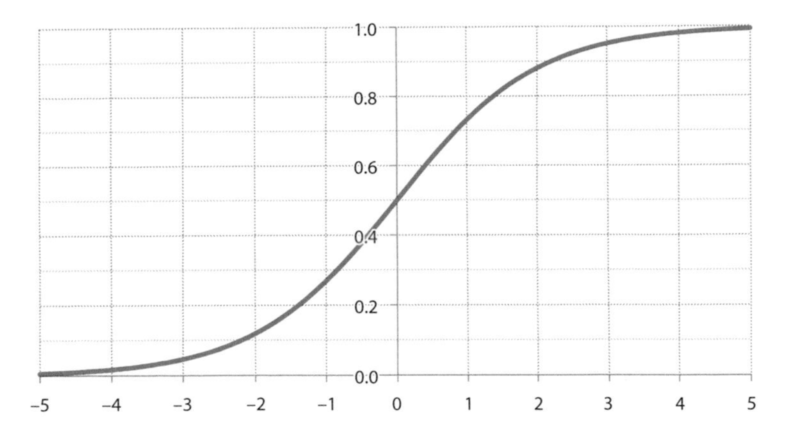

FIGURE 4.16. Example of a logistic curve with $A = C = 1$, $B = 1/e \approx 0.368$.

S-shaped, with rapid growth for a time, before eventual stabilization as the high score gets harder and harder to top. Similar behavior also arises in the study of population growth, the adoption of new technologies, and the growth of tumors; in each case, logistic curves serve as useful predictive models. Such models have also been used in some of the sports contexts mentioned before.

Mathematically, a logistic equation is any equation of the form

$$F(x) = \frac{A}{1 + C \times B^x}$$

for some parameter A and positive parameters B and C.

Since we've got some exponential growth going on in the denominator, the maximum value of $F(x)$ is A, and the minimum value is 0. When $0 < B < 1$, the curve looks like an S, with $F(x)$ tending to A as x tends toward ∞, and $F(x)$ tending toward 0 as x tends toward $-\infty$ (Figure 4.16). When $B > 1$, the curve looks like a Z and the opposite is true: F gets closer to A as you move to the left, and it gets closer to 0 as you move to the right.

By adding a vertical shift parameter, one can also consider a family of shifted logistic equations:

$$F(x) = \frac{A}{1 + C \times B^x} + D.$$

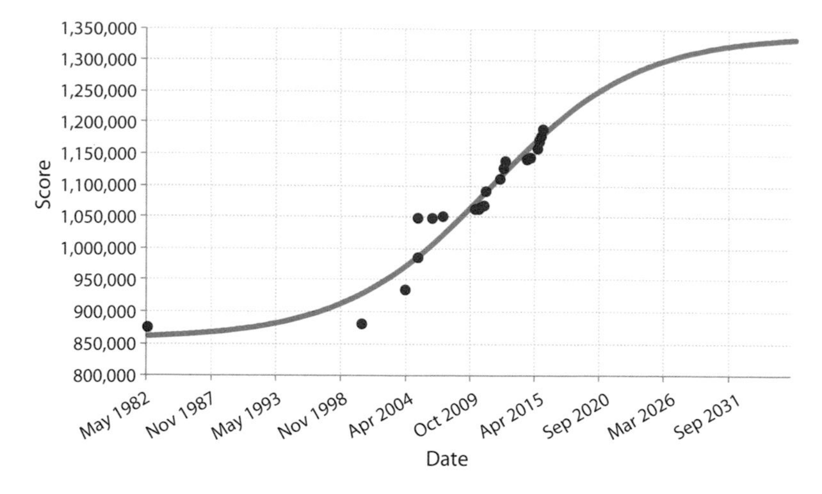

FIGURE 4.17. A logistic regression curve for the *Donkey Kong* high-score data. The data point on the far left corresponds to Billy Mitchell's original high score.

The difference here is that the curve no longer needs to tend to 0 on one side of the graph. For example, if $0 < B < 1$, then F tends to D as x tends to $-\infty$, and F tends to $A + D$ as x tends to ∞.

As it turns out, a shifted logistic model fits the *Donkey Kong* high-score data pretty well. The parameters of the logistic curve used in Figure 4.17 are given by $A \approx 486{,}747$, $B \approx 0.999563$, $C \approx 103.102$, and $D \approx 856{,}573$.[23] According to this model, there's still room for significant improvement: the predicted maximum score is more than 100,000 points higher than the current record! And if you want that high score all for yourself, it looks like you've got some time, too: the way things are currently going, it'll be several years before the high score breaks 1.3 million points.

Of course, this isn't the only approach one could use to study the limits of high scores in our favorite games. In fact, other people have studied *Donkey Kong* in order to estimate the maximum possible score. One online discussion analyzed the scoring rules of *Donkey Kong* and determined that the highest score should be somewhere around 1,301,300 points, which isn't that far off the estimate provided here (the model predicts a maximum closer to 1,343,000).[24]

The benefit to the logistic approach, however, is that it doesn't depend on the scoring rules of the game. You could fit a logistic curve to any set of high-score data. But in order to have enough data on when high scores have been broken, you need a game popular enough that people care about having the highest score. For a game like *Donkey Kong*, this isn't a problem, but for more obscure titles, it could be.

4.9 CLOSING REMARKS

There is some beautiful mathematics hidden in the scoring rules for different games. In our modern and interconnected world, this even has some practical application, as a deep understanding of a game's scoring mechanics can help determine whether a posted high score could have been obtained dishonestly. We've also seen how we can use math to optimize your strategy for getting an ideal score or to predict the future of high scores for classic games.

The moral here is that scoring systems always involve some mathematics. But certain games, either through their scoring rules or through their popularity, invite deeper mathematical inquiry than others. So the next time you're playing a game with a scoring mechanic, take a moment to meditate on its scoring rules. If you look at things from the right perspective, you may find some rich mathematics lurking just beneath the surface.

5

The Thrill of the Chase

5.1 I'MA GONNA WIN!

When the Nintendo 64 was released stateside in the fall of 1996, there were only two games available. In fact, the number of games didn't reach double digits until almost four months later, when *Mario Kart 64* hit the shelves. Nintendo's turn-of-the-millennium console was never known for its large library of games,[1] but what the N64 may have lacked in quantity, it made up for in quality. Games like *Mario Kart 64*, *GoldenEye 007*, and *The Legend of Zelda: Ocarina of Time* are now recognized by many as classics of their generation.

Mario Kart 64 in particular still maintains a devoted following over twenty years after its original release. The multiplayer is unmatched, even by later entries in the series, and world rankings for the best players are still updated at mariokart64.com. Though the game hasn't aged that well graphically, from a gameplay perspective, it's still top-notch.

As the name suggests, games in the *Mario Kart* series feature Mario (along with his friends, enemies, and frenemies) in go-karts. Unsurprisingly, the goal is to reach the finish line before your competitors. What sets the *Mario Kart* series apart is its use of special items to enhance gameplay. These items range from turtle shells[2] you can shoot at other karts to mushrooms that boost your speed. If you're in last place, you can use items to pursue the other racers; if you're in first place, you can use them to avoid pursuit and hold onto your lead. There are few things more satisfying than shooting from last place to first all because of a particularly lucky draw from the item box.

In many games—not just *Mario Kart*—this tension between the pursuer and the pursued makes for compelling gameplay. Indeed, many of our favorite pastimes involve some type of dynamic between these two forces. In football (American or otherwise), the player in control of the ball needs to evade pursuit from members of the opposing team. The same is true in many other goal sports, such as basketball and hockey. Even the simplest childhood games involve some form of chase: think of duck-duck-goose or tag.

It's no surprise, then, that such a fundamental dynamic also emerges in many video games. Whether you're trying to catch up to a racer ahead of you, evade a fighter pilot behind you, or hunt down an opponent in a good old-fashioned foot chase, the roles of the hunter and the hunted emerge in so many games that it would be nearly impossible to list them all.

The mathematics of pursuit and evasion is similarly rich. A number of questions fall under this umbrella, at varying levels of sophistication and intersecting a wide assortment of mathematical topics, including geometry, game theory, and differential equations. Rather than highlighting all the possible connections between pursuit and evasion problems and video games, I'd like to explore just a handful of the most interesting examples. Of course, what makes an example interesting is a matter of taste; the examples we'll consider here are ones where the mathematics is particularly rich and the game is particularly important within the larger history of games.

As you might expect, *Mario Kart* poses some interesting questions that can be explored within this context. Let's begin with a foray into the Mushroom Kingdom, and then look at two other interesting examples of pursuit and evasion in earlier video games.

5.2 SHELL GAMES

Every entry in the *Mario Kart* series offers up an alternative to racing, known as battle mode. In this mode, players use items to attack others and defend themselves. Players start with three balloons attached to their karts. Every hit you take costs you a balloon, and the last player with balloons left is the winner.

Just as among thieves, there's no honor among players in battle mode. It's a lawless land where alliances are made and broken in an instant. But, speaking as someone who has spent more time in said land than he'd care to admit, it's also a lot of fun. More to the point, it's also

FIGURE 5.1. A bird's-eye view of the Block Fort.

a place where interesting questions of pursuit and evasion come to the forefront.

Specific battle courses are designed for each entry in the series, but few of them compare to the Block Fort course from *Mario Kart 64*. Beautiful in its simplicity, the course takes place on a square grid, with four symmetrically designed buildings at each corner (Figure 5.1). Drivers can travel between buildings using aboveground bridges and can also chase each other through the streets between buildings.

There are a number of items available to players in battle mode, though the specifics vary from game to game. Two items, however, have appeared in every game without fail, and they provide a perfect entry point into the study of pursuit and evasion.

The items we focus on here are green shells and red shells. Both are kid-friendly analogues of missiles, but their behavior is quite different. Green shells follow a straight trajectory once fired, mindlessly bouncing off of walls until they hit a target or the game ends. Red shells, on the other hand, are heat-seeking. They lock onto the nearest opponent and engage in a targeted pursuit. Unlike green shells, however, they can't bounce off walls. If they strike a wall, they break apart and no longer pose a threat. Therefore, with some evasive maneuvering it's possible to avoid getting hit by a red shell headed your way.

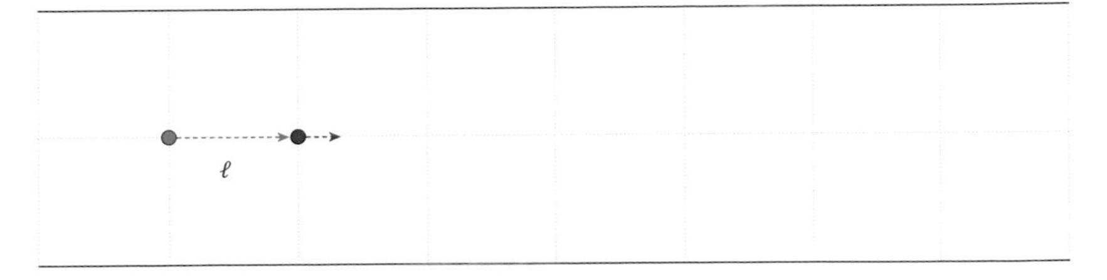

FIGURE 5.2. Your shell is represented by the green dot. Your enemy is blue. The distance between your initial position and your enemy's initial position is what we're calling ℓ.

When you're in pursuit of another player and pick up one of these items, what should you do? Where should you fire the shell, and when? Intuitively, you can probably get a good handle on these questions by playing the game, but we can also take a systematic approach. Let's study the behavior of these weapons on certain areas of the Block Fort course. The mathematics can quickly become intractable if we're not careful, so we'll keep the modeling relatively simple. You'll find that there are several ways to generalize what we do here, should you be so inclined.

5.3 GREEN-SHELLED MONSTERS

Since green shells have simpler behavior, let's investigate them first. Suppose you're carrying a green shell and chasing someone along a straight portion of the Block Fort course. For now, let's assume that you and your target are both in the middle of the road and traveling straight ahead.

When you fire at your opponent, your shell zips along with some fixed speed. Let's call it v_S (S for shell). Similarly, let's assume that your opponent's kart moves with speed v_K (you guessed it, K for kart). The shell will never catch up if its speed is less than the kart's, so let's assume that $v_S < v_K$. Lastly, let's suppose that your opponent initially has a head start of length ℓ.

The most obvious thing to do is simply fire the shell straight ahead. Assuming that your opponent doesn't change directions, your shell hits its target when the distance it has traveled from your kart is equal to the distance your enemy has traveled from his starting point, plus ℓ (Figure 5.2). In other words, if t denotes the time until the collision between shell and kart, we must have

$$v_S t = v_K t + \ell.$$

Solving for t tells us how long it takes for the shell to find its target:

$$t = \frac{\ell}{v_S - v_K}.$$

All the features of this equation agree with our intuition. First, since $v_S > v_K$, the collision time is always positive, though it gets larger the closer v_S is to v_K. As the attacker, however, you have no control over v_S or v_K. Also, t decreases as ℓ decreases; in other words, the closer you are when you fire the shell, the sooner it hits your opponent. Indeed, one of the most effective ways to use a green shell is to get right behind an opponent before firing.

While this is the most direct approach, it's also the easiest to avoid. Provided your opponent knows you're coming and hears the shell fire, movement to the left or right easily dodges the shot, especially if you're firing from far away.

Moreover, some items have defensive capabilities that would make a direct attack ineffective. In *Mario Kart 64*, for instance, players can keep red shells, green shells, and bananas, all of which can be used defensively. By holding down the trigger button, each of these items sits at the tail of the player's kart until the player is ready to fire. During this time, if a green or red shell hits the player from behind, the item takes the impact instead. The player is left unharmed.

In order for a green shell to be effective when your opponent is using an item defensively, then, you'll need to bank a shot off one of the walls of the course. In so doing, you'll be able to hit your opponent from the side, rather than from behind the kart. This is where most players simply fire at random and hope for the best. Such a strategy, in addition to being not terribly effective, also comes with a cost: if you don't think at least a little bit about the trajectory of the shell, there's a risk that it could ricochet off a few walls and hit you!

So, let's take a more methodical approach and explore the mathematics involved in modeling a banked shot. There are a number of ways to do this, but we'll start with the simplest case: bouncing a shell off of one wall before hitting the other kart.

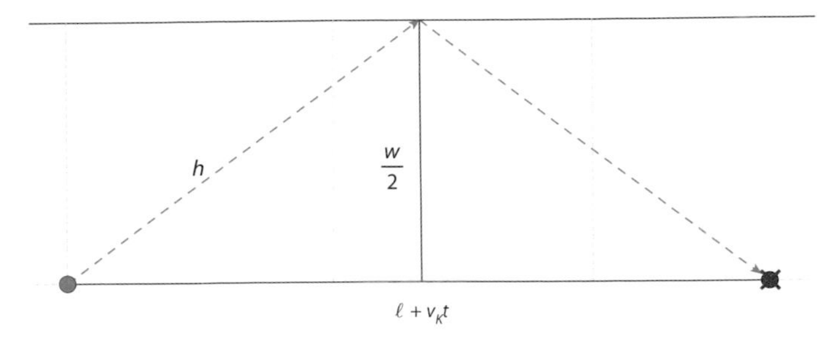

FIGURE 5.3. Annotated trajectory with one bounce.

In this case, we'll need to introduce another variable for the width of the road. We'll call this width w. We now want to find a path to our target that looks like the one in Figure 5.3.

How does this change the time to impact, and at what angle should you fire your shell? To answer these questions, we'll need to use a bit of geometry. As before, the distance your opponent travels equals $\ell + v_K t$. Your shell, however, must now travel a greater distance. Splitting the shell's trajectory into two linear segments of length h, we see that $2h = v_S t$.

This isn't quite enough to determine the new value of t, however, since we don't know the size of h. Thankfully, the Pythagorean theorem provides us with an additional piece of information! From Figure 5.3, we see that h is the length of the hypotenuse in a triangle with sides of length $\frac{w}{2}$ and $\frac{\ell+v_K t}{2}$. This means that

$$\left(\frac{w}{2}\right)^2 + \left(\frac{\ell + v_K t}{2}\right)^2 = h^2.$$

On the other hand, we already know that $4h^2 = (2h)^2 = (v_S t)^2$. Combining this with the equality above, we find that

$$(v_S t)^2 = w^2 + (\ell + v_K t)^2 .$$

This is a quadratic equation in t; by rearranging terms, we can rewrite the above equality as

$$\left(v_S^2 - v_K^2\right) t^2 - 2\ell v_K t - w^2 - \ell^2 = 0.$$

The quadratic formula provides us with the solutions to this equation. As you may recall, in general the formula provides us with two solutions.

In this case, those solutions are of the form

$$t = \frac{\ell v_K \pm \sqrt{\left(v_S^2 - v_K^2\right) w^2 + (v_S \ell)^2}}{v_S^2 - v_K^2}.$$

However, since our solution must be positive, we must take the positive value of the square root.[3] In other words, if you bank your shot off one of the walls, the time required to hit your opponent is now

$$t = \frac{\ell v_K + \sqrt{\left(v_S^2 - v_K^2\right) w^2 + (v_S \ell)^2}}{v_S^2 - v_K^2}.$$

Not nearly as simple as before, unfortunately. However, if $w = 0$, the above formula reduces to the one we derived for a straight trajectory, evidence that this is indeed the formula we're looking for. Moreover, if $w > 0$, this expression tells us that it takes longer for your shell to reach its target if you bank the shot than if you fire it directly, which makes sense since it has to travel farther.

Of course, in order for the shell to hit its target, you need to know where to fire it. Since you're not shooting directly at your target, this can make a successful bank shot difficult. With what we now know, however, we can also come up with an expression for the angle your shell should make with the horizontal.

Let's call this angle θ. From our diagram above, the cosine of this angle is given by

$$\cos\theta = \frac{\ell + v_K t}{2h} = \frac{\ell + v_K t}{v_S t} = \frac{\ell}{v_S t} + \frac{v_K}{v_S}.$$

In other words,

$$\begin{aligned}\theta &= \cos^{-1}\left(\frac{\ell}{v_S t} + \frac{v_K}{v_S}\right) \\ &= \cos^{-1}\left(\frac{w^2}{w^2 + \ell^2}\left[\frac{v_K}{v_S} + \frac{\ell}{w}\sqrt{1 + \left(\frac{\ell}{w}\right)^2 - \left(\frac{v_K}{v_S}\right)^2}\right]\right).\end{aligned}$$

Just as with the simpler example, the only variable that you have control over here is ℓ.[4] The width of the Block Fort track is fixed, as are the speeds of the kart and the shell. And while it's unlikely you'll be measuring any of these parameters in the heat of battle, we can still use the above formula to come up with some useful heuristics.

In *Mario Kart 64*, for example, the ratio v_S/v_K is around 1.5, with some variation due to difficulty and kart driver. So, if you were driving behind someone whose distance to you was around twice the width of the track, this formula tells you that for a bank shot off the wall, your firing angle should be about $\cos^{-1}(0.987) \approx 9.2^\circ$.[5] In other words, it doesn't take a huge angle to make a successful bank shot.

5.4 GENERALIZATIONS AND LIMITATIONS

So far, we've only looked at two different shooting strategies: attacking head-on or banking your shot off of a single wall. But there's no reason to restrict our attention to a bank shot involving just one bounce. In fact, for the particularly savvy player, everything we did in the last section can be generalized to the case of m wall bounces. Conceptually, this doesn't require any new ideas, so I'll leave the details for curious readers to fill in (it basically amounts to inserting m at appropriate places). The end result is that if you want to bounce your shell off the wall m times before hitting your target, the time it will take, which we'll call t_m, is given by

$$t_m = \frac{\ell v_K + \sqrt{\left(v_S^2 - v_K^2\right) m^2 w^2 + (v_S \ell)^2}}{v_S^2 - v_K^2}.$$

Similarly, you should fire the shell at an angle of

$$\begin{aligned} \theta_m &= \cos^{-1}\left(\frac{\ell}{v_S t_m} + \frac{v_K}{v_S}\right) \\ &= \cos^{-1}\left(\frac{1}{(mw)^2 + \ell^2}\left[\frac{v_K}{v_S}(mw)^2 + \ell\sqrt{(mw)^2 + \ell^2 - \left(mw\frac{v_K}{v_S}\right)^2}\right]\right). \end{aligned}$$

Notice that when $m = 0$ or 1, these formulas agree with the ones we've already derived. Figure 5.4 shows some images of trajectories using our original example with $m = 2$, 3, and 4.

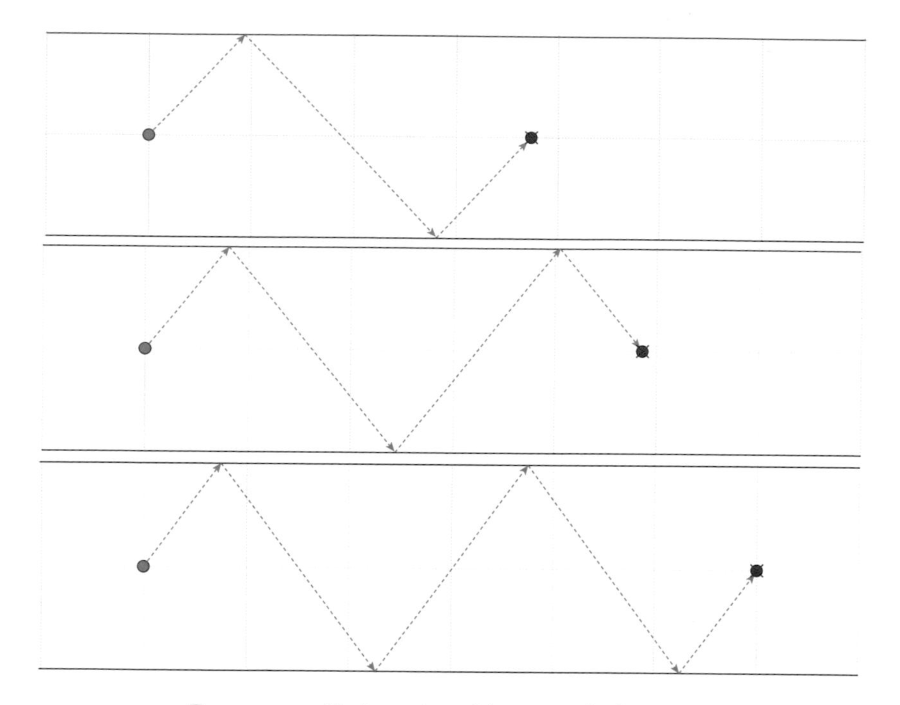

FIGURE 5.4. Trajectories with 2, 3, and 4 bounces.

Looking at these trajectories raises some additional interesting questions. Here are a few (with answers for the impatient in the footnotes):

- What happens to the angle θ_m as m increases? Does it ever reach 90° or does it plateau earlier?[6]
- How much farther can the opponent's kart go with each increase in the value of m? Based on this, approximately how many of these trajectories are actually viable?[7]

There are other ways to generalize as well. You can explore what happens if your trajectory (or that of your opponent) isn't perfectly horizontal. Or you can keep your paths horizontal, but move one or both of you off the center of the road.

Of course, one of the most practical ways to generalize would be to consider a different segment of the course or the entire course at once. Unfortunately, the price of generality in such a model probably isn't worth the loss in simplicity. In general, if you've got a green shell, it's not worth strategizing *too* much. Since it's tough to calculate angles in

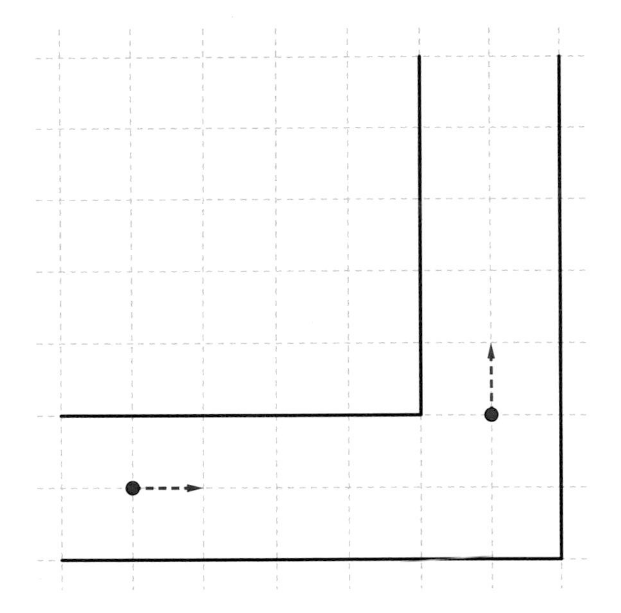

FIGURE 5.5. Chasing an opponent (blue) near a corner.

the heat of battle, and you won't be able to anticipate your opponents' future moves, it's probably best to fire at close range or hunt for some better items.[8]

5.5 SEEING RED

Now let's suppose you're chasing someone who rounds a corner and disappears from your sight (Figure 5.5). While you could certainly analyze this scenario as we did before, by trying to bank a green shell off the walls of the course, this quickly begins to feel overly complicated. The best strategy if you have a green shell may be simply to pick a direction and fire away.

But what if you have a red shell? In this case, it locks onto your target once it's fired. It may seem like there's no strategy involved here, but that's not quite true. If you're too far behind, the shell may round the corner too early and hit the side of the course instead of your intended target. In other words, fire too soon and your red shell may go to waste.

But how soon is too soon? Before getting into the mathematical details, let's take a look at some examples of successful and unsuccessful shots.

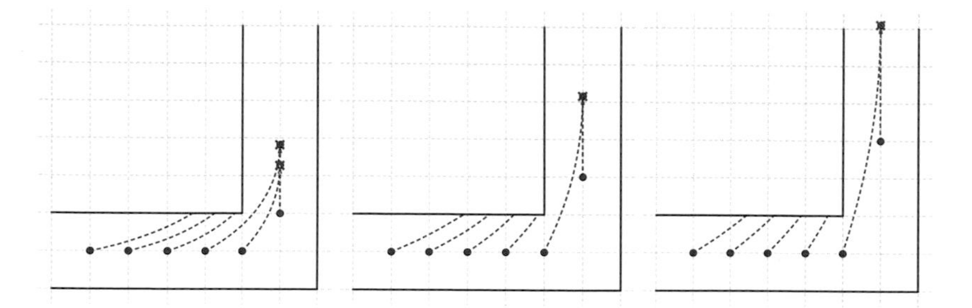

FIGURE 5.6. Trajectory of a red shell. If you're too far from the corner, your shell is more likely to hit the wall.

There are two ways we can explore how the trajectory of your shell varies. The most obvious is to fix the velocity of your shell but vary the starting position of your shell and your opponent's kart. Figure 5.6 shows some sample trajectories assuming that your shell is twice as fast as your opponent's kart (the addendum to this chapter has details on how these curves are derived).

A less obvious (though still helpful) approach is to fix the position of the shell and kart but vary the velocity of the shell. This is illustrated in Figure 5.7. Note that if the shell is very fast, it can round the corner successfully. Too slow, though, and your opponent can escape unscathed.

Of course, this isn't terribly practical, since you don't have control over how fast the shell travels. This analysis does, however, provide insight into a *necessary* condition for your red shell to hit its target.

Notice that the bottom trajectory is a straight line. This line corresponds to the case where your shell moves infinitely fast and hits your opponent instantaneously.[9] Every other trajectory moves closer and closer to the wall. Therefore, if that straight-line trajectory hits the wall, so must the actual trajectory.

So where does the straight-line trajectory hit the wall? Let's label your opponent's kart starting position as $(0, \ell_K)$ and your shell's starting position as $(-\ell_S, 0)$, for some positive values ℓ_S and ℓ_K. (We're taking the origin of our coordinate system to be right in the middle of the track's corner.) Then the segment joining these two points lies on the line described by the equation

$$y = \frac{\ell_K}{\ell_S}x + \ell_K. \tag{5.1}$$

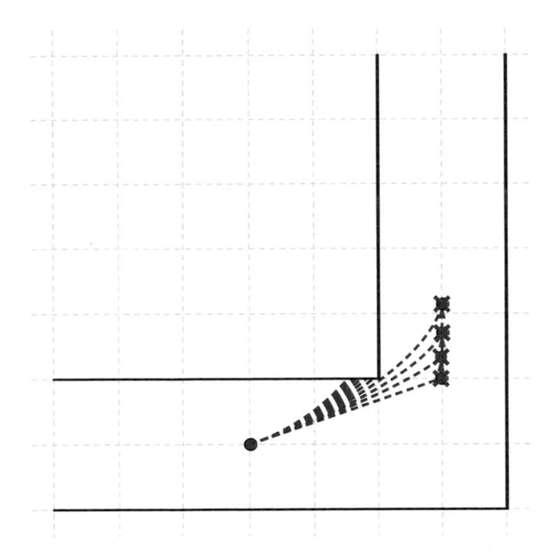

FIGURE 5.7. Trajectories for the red shell with fixed starting positions and varying speeds. The higher up impact occurs, the slower the red shell.

Based on the diagram in Figure 5.8, when $x = -w/2$, the straight-line trajectory intersects the wall if $y > w/2$, and does not intersect the wall otherwise. In other words, it's necessary for

$$y = \frac{\ell_S}{\ell_K}x + \ell_S = -\frac{\ell_S w}{2\ell_K} + \ell_S < \frac{w}{2}.$$

Rearranging terms, this is equivalent to the inequality

$$\frac{2}{w} < \frac{1}{\ell_K} + \frac{1}{\ell_S}. \tag{5.2}$$

Notice that the right side decreases as ℓ_K and ℓ_S increase, which makes sense—the farther either party is from the corner, the more likely the shell is to meet a premature demise.

Of course, while this is a *necessary* condition, it's not a *sufficient* one. Just because the *fastest* trajectory would hit the shell, this doesn't mean the *actual* trajectory will: if your shell is relatively slow, it still might hit the wall. If you're interested in finding a more precise inequality, including a description of the path traced out by the red shell, check out the addendum to this chapter. For now, let's move on to the next example of pursuit and evasion that's worth, well, pursuing.

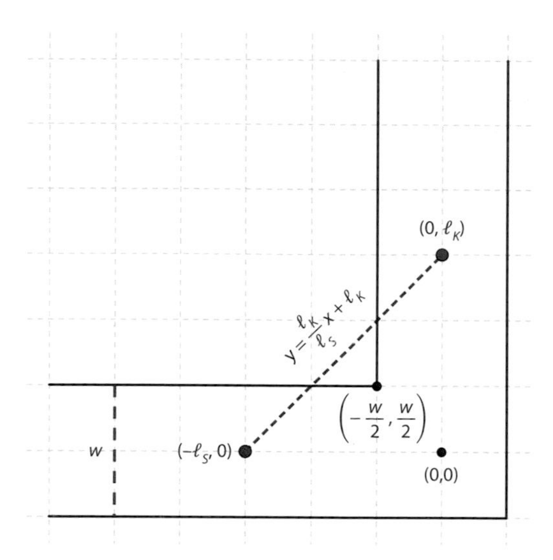

FIGURE 5.8. Labeled diagram of our model. The straight line represents the trajectory of a shell with infinite speed. When the line crosses the course wall, your shell will never hit its target. Note that the origin of our coordinate system is right in the middle of the lower-right corner.

5.6 APOLLONIUS CIRCLE PURSUIT

The *Mario Kart* series is hardly the only one in which pursuit and evasion play a role. In fact, interesting mathematics emerges in some of the earliest video games. For our next stop on the pursuit and evasion tour, let's return to a time when Atari reigned supreme.

In 1980, Atari released the arcade game *Missile Command*. In the game, players must protect six cities from an onslaught of ballistic missiles by controlling three anti-missile batteries. The player moves a crosshair on the screen and presses fire to launch a counter-missile; the battery closest to the crosshair then fires.

Releasing a game about missile defense in 1980 was certainly topical, and this may have contributed to *Missile Command*'s success. Today, the game is still touted as one of Atari's classics, and it has been rereleased many times (with revamped graphics, gameplay, or both). Some version of the game—either a genuine remake, or a clone attempting to piggyback on the original game's success—can be found on every game console and in every app store. In fact, you can even play a version of the game online for free at Atari's website.

FIGURE 5.9. The red lines represent the ballistic missiles attacking the six cities below. The missile batteries are on the left, the right, and in the middle. Screenshot from *Missile Command*, reproduced with permission from Atari.

Initially, the game is fairly forgiving. Missiles appear infrequently and move slowly, giving the player plenty of time to line up the perfect shot. In later levels, though, the missiles become faster and more numerous. Some missiles also split into multiple missiles if you don't take them out quickly enough. Because of this, it's important to try to destroy missiles as soon as you see them.

This raises an important question. As missiles get faster, it gets harder to know where to fire. After the first few levels, it's pointless to put your crosshair right on the head of a moving target, since by the time your counter-missile reaches its destination, the missile you were trying to hit will have moved away. Given a missile's initial location and trajectory, then, where should the player position her crosshair to ensure that the counter-missile will hit its target?

Before we can answer this question, we need to define some variables. Clearly the speeds involved are important: if the player's missile speed is much higher than the enemy's, the game is easy, and if it is much lower, the game is hard. Let v_E denote the enemy's missile speed, and v_P denote the player's.[10] As it turns out, we are most interested in the ratio of these speeds, so to simplify notation we define k to be v_E/v_P. In particular, when $k < 1$ the player's missile is faster than the enemy's, and vice versa when $k > 1$. Note that in the game, $k < 1$ initially, but as the game progresses the value of k increases.

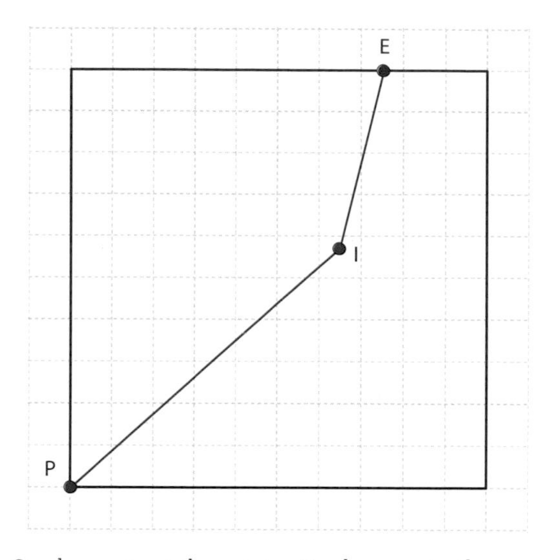

FIGURE 5.10. Our base sits at the origin, P, when a missile enters the screen at position E. Does an interception point I exist?

Finally, we need some coordinates. To simplify the analysis, we focus on only a single missile battery and a single inbound enemy missile. We also assume that the missile is coming from the right of the battery, so that our analysis takes place only in the first quadrant of the coordinate plane; this really isn't so limiting, since a missile coming in from the left would just require some switching of signs in our coordinates. Finally, we take the origin of our system to be the location of the battery, which we label P. Similarly, we label the enemy missile's initial position E. Now the question becomes: Is there a point I where we can intercept the enemy's missile? Figure 5.10 provides an example of what we're looking for.

5.7 OVERVIEW OF A WINNING STRATEGY

Our goal now is to figure out whether such a point I exists, and if so, where it is. As a first step, though, let's ignore the *direction* of the incoming missile and consider only its initial *position*. Given only the speeds of the two missiles, we can then start with a simpler question: what's the set of all points where the missiles *could possibly* meet? We denote this set by S.

The key to determining S is the fact that the distances traveled by the two missiles are related to one another. More specifically, the ratio of the distances traveled by the missiles when they meet at a point on S is equal to the ratio of their speeds. In other words, in order for the missiles to hit at a point I after a time t, it must be true that

$$\frac{IE}{IP} = \frac{v_E t}{v_P t} = k.$$

A point I is in S if and only if the above equality holds. Using a little coordinate geometry, we can use this to determine S. Since P is at the origin, its coordinates are (0, 0); for the enemy's location, we set $E = (E_1, E_2)$, and we write the (to be determined) coordinates of I as $I = (x, y)$. Using the distance formula, we have

$$k = \frac{IE}{IP} = \frac{\sqrt{(x - E_1)^2 + (y - E_2)^2}}{\sqrt{x^2 + y^2}}.$$

With a little bit of algebra (including completing the square in x and y), the above equation can be rewritten as

$$\left(x - \frac{E_1}{1 - k^2}\right)^2 + \left(y - \frac{E_2}{1 - k^2}\right)^2 = \left(\frac{\|E\|}{k^{-1} - k}\right)^2, \tag{5.3}$$

where $\|E\| = \sqrt{E_1^2 + E_2^2}$ is the distance between the enemy's missile and the player's battery.

The equation above is of the form $(x - a)^2 + (y - b)^2 = r^2$, which, describes a circle of radius r and center at (a, b)! In other words, the set S, defined to be the points where the two missiles could potentially meet, is a circle. This circle is called the Apollonius circle of P and E.

Of course, to get a sense for where this circle lies in our coordinate plane, we could plot some examples for different values of E and k. There are, however, some general properties worth pointing out. First, since the center of S is $\left(\frac{E_1}{1-k^2}, \frac{E_2}{1-k^2}\right)$, the distance between P and the center of S is equal to

$$\frac{\|E\|}{|1 - k^2|} = \frac{r}{k},$$

where $r = \frac{\|E\|}{|k^{-1} - k|}$ is the radius of S. In particular, P is inside the circle S if and only if $k > 1$. A similar argument shows that E is inside the circle

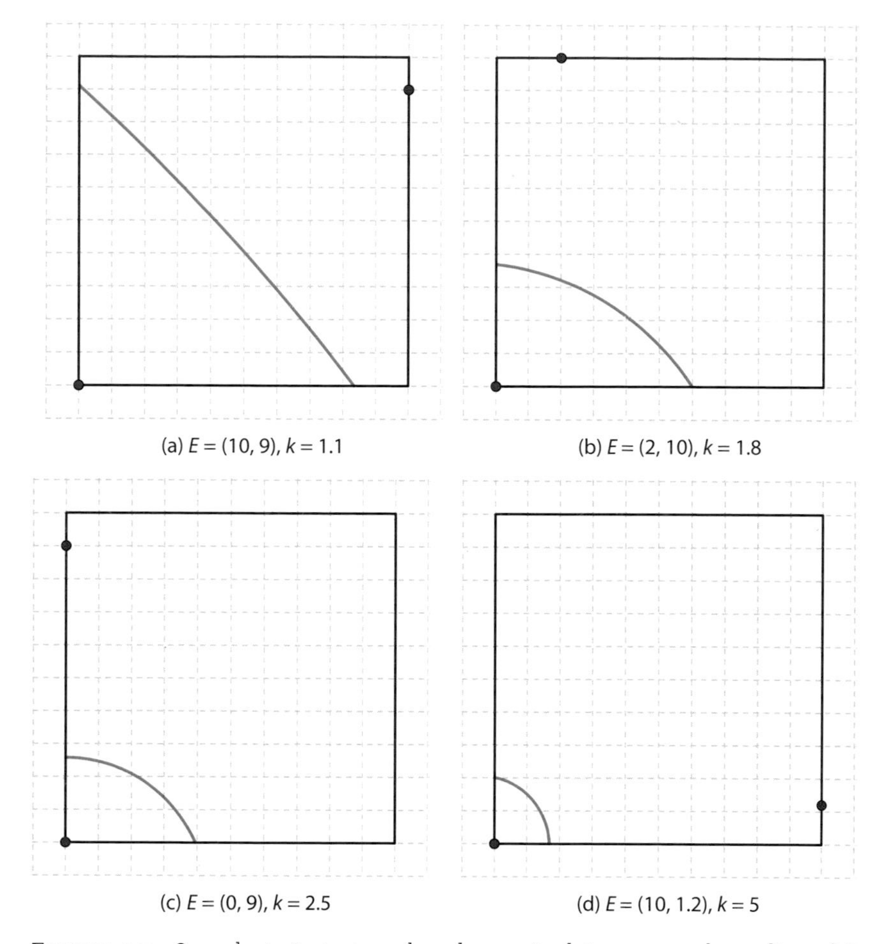

FIGURE 5.11. Sample trajectories when $k > 1$. As k increases, the radius of S decreases. In every case, interception is possible if and only if the endpoint of the missile's trajectory lies inside the circle S.

S if and only if $k < 1$. Of course, the equation defining S isn't valid when $k = 1$; in this case, the mathematics simplifies quite a bit (as you should verify!).

Which of these two scenarios do you think is worse for the player? You may guess that the case $k > 1$ should be more restrictive, since this corresponds to the case when the enemy's missile is faster than the player's, thereby making the game more difficult. Let's begin with this case. Take a look at the examples in Figure 5.11 with varying E and k when $k > 1$.

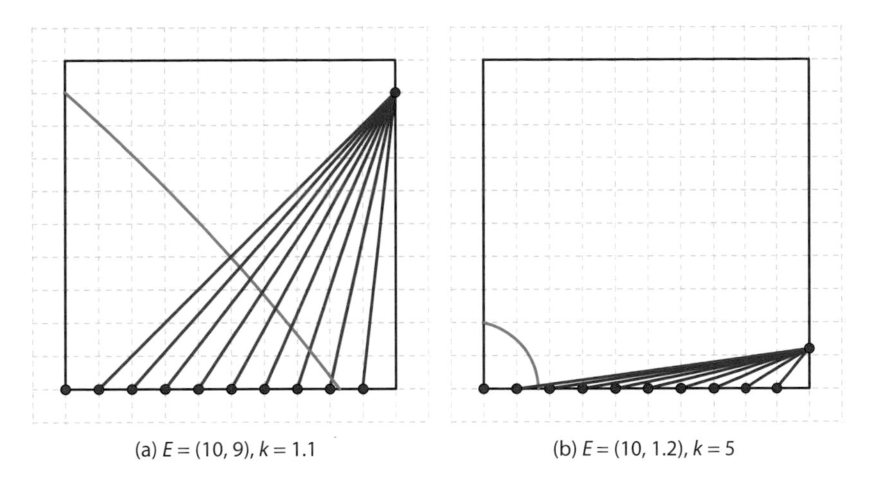

FIGURE 5.12. In these images, blue trajectories are ones you can intercept (i.e., ones that intersect S), while red trajectories are impossible to stop. On the left, k is close to 1, so S has a larger radius and it's more likely you'll be able to intercept the missile. On the right, k is large, so S has a smaller radius and most missiles are impossible to intercept.

In each example, the green arc represents the part of S that lies in the first quadrant. The location of S helps us determine whether or not we'll be able to intercept the enemy's missile. More specifically, if the missile's trajectory intersects S, then we'll be able to intercept it. Otherwise, we won't be able to. The situation gets worse as k increases: the larger its value, the smaller the arc and the fewer trajectories we'll be able to intercept. For some visual examples of this phenomenon, see Figure 5.12.

We can come up with a fairly simple sufficient condition for interception. Since the origin P is inside the circle S when $k > 1$, it must be true that S intersects the positive x-axis. Let's call this intersection $(S_0, 0)$ for now (we'll get more precise later). Similarly, let's denote the destination of the enemy's missile by the point $(D, 0)$—the missile always targets a point on the x-axis, since that's where the cities and batteries are. In order for there to be an intersection point between S and the missile's trajectory, we must have

$$D < S_0.$$

We'll return to this inequality in a moment.

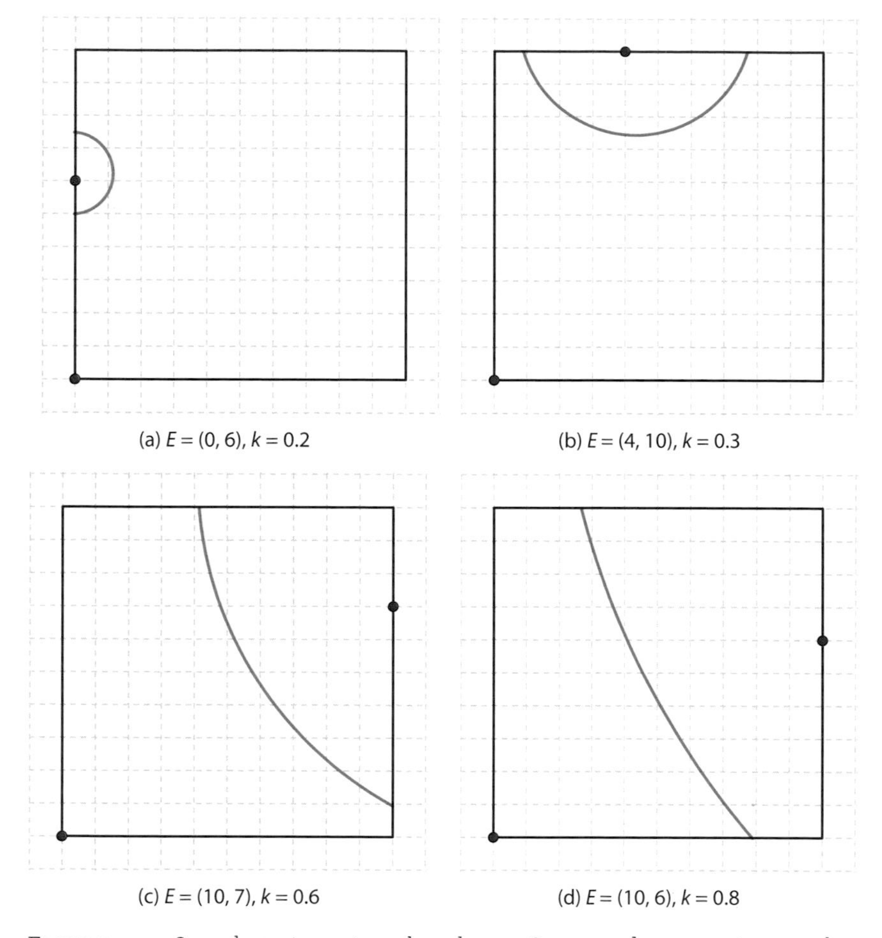

FIGURE 5.13. Sample trajectories when $k < 1$. Can you draw a trajectory that can't be intercepted?

Next, let's consider what happens when $k < 1$. It's tempting to say that if your missile is faster than your enemy's, you should always be able to intercept it. After all, in this case the enemy missile always starts inside of S, so at some point it must cross the boundary of S, right?

To help develop our intuition, take a look at the examples in Figure 5.13. In these examples, smaller circles correspond to values of k near 0, and larger circles correspond to values of k near 1. In the first three examples, it's possible to intercept any missile, because any line segment from the missile's starting position to a target on the x-axis

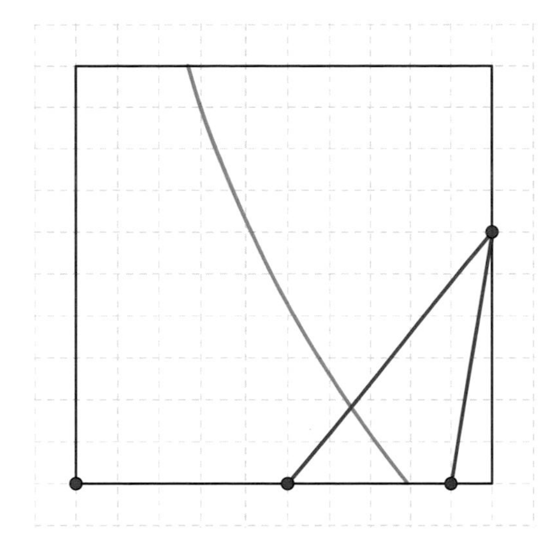

FIGURE 5.14. Two possible trajectories when k is close to and less than 1.

must go through S. This isn't necessarily the case for the fourth image, though. Do you see what can go wrong?

Take a look at the two trajectories in Figure 5.14. You can intercept the blue trajectory without a problem, since it passes through S. However, when a missile enters from the right-hand side of the grid and its final target is also on the right-hand side of the grid, interception may not be possible.

More precisely, when $k < 1$, the only way for S not to intersect the trajectory of the enemy's missile is if the missile enters from the right and intersects the x-axis to the right of S. In other words, we once again need D to be less than S_0.

Unlike the previous case, however, there's no guarantee that S_0 exists, i.e., that S intersects with the positive x-axis. In fact, in three of the four $k < 1$ examples shown above, it doesn't! When S doesn't intersect the x-axis, we're always guaranteed to find a suitable intersection point I.

5.8 PINPOINTING THE INTERSECTIONS

The possible location of I is less interesting than the conditions under which it exists, so let's consider the latter issue before the former. From what we've done so far, we know that the intersection question naturally splits into two cases:

1. S doesn't intersect the x-axis. In this case, $k < 1$ and I is guaranteed to exist.
2. S intersects the x-axis. In this case, I exists if and only if $D < S_0$.

Let's start with the first case. When S doesn't intersect the x-axis, this means that $(x, 0)$ can never be a solution of equation (5.3), the equation defining S. Setting $y = 0$ and simplifying, this is equivalent to the statement that

$$x^2 - \frac{2E_1}{1-k^2}x + \frac{\|E\|^2}{1-k^2} \tag{5.4}$$

can't equal zero.

The above expression is a quadratic in x. Again recalling our high school algebra, the only way for an expression of the form

$$ax^2 + bx + c$$

to not have any real zeros is for the discriminant of the equation to be negative, i.e.,

$$b^2 - 4ac < 0$$

(this is a consequence of the quadratic formula, which is rearing its head yet again).

Returning to *Missile Command*, this argument tells us that if S doesn't intersect the x-axis, it must be the case that

$$\left(\frac{E_1}{1-k^2}\right)^2 - \frac{\|E\|^2}{1-k^2} < 0.$$

After simplification, we see that the above inequality is equivalent to

$$k < \frac{E_2}{\|E\|}. \tag{5.5}$$

This delightfully simple inequality confirms the evidence provided by our earlier images: indeed, the condition that $k < 1$ isn't sufficient to guarantee we'll be able to intercept enemy missiles. If E_2 is very small (that is, if the enemy's missile enters the screen very low to the ground), we may not be able to prevent an attack even if our missiles are fast.

Note that $\|E\| = \sqrt{E_1^2 + E_2^2} > E_2$, so inequality (5.5) is stronger than the inequality $k < 1$. In other words, to guarantee an intersection, it's not enough for our missiles to be faster than those of our enemies; they have to be faster by an amount that depends upon the enemy missile's location.

Now let's turn our attention to the second case, when the circle S intersects with the x-axis. This means that expression (5.4) must have at least one real root. By the quadratic formula, the roots of expression (5.4) are given by

$$\frac{E_1 \pm \sqrt{(k\,\|E\|)^2 - E_2^2}}{1 - k^2}.$$

We know that in order for an intersection to exist, we must have $D < S_0$, where S_0 is one of these roots. But which one?

Well, if $k > 1$, the circle S contains the origin, and so one of the roots is positive and one is negative.[11] In this case, the positive root is the one we're interested in. On the other hand, if $E_2/\,\|E\| < k < 1$, then both roots are positive (prove it!) and it's the smaller root we're interested in, since the larger one is outside of the game screen. In this case, we take the value with the negative square root.

Bringing this all together, we can say that it's possible to intercept the enemy's missile if one of the following three conditions holds:

(i) $k < E_2/\,\|E\|$,
(ii) $E_2/\,\|E\| < k < 1$ and $D < \frac{E_1 - \sqrt{(k\|E\|)^2 - E_2^2}}{1-k^2}$,
(iii) $k > 1$ and $D < \frac{E_1 + \sqrt{(k\|E\|)^2 - E_2^2}}{1-k^2}$.

Once you know that an intersection exists, finding it is a matter of calculating the point of intersection between a circle and a line. The equation of the relevant circle is given by equation (5.3). Meanwhile, since the trajectory of the enemy's missile is a line passing through $(D, 0)$ and (E_1, E_2), it has equation

$$y(E_1 - D) = E_2\,(x - D)\,. \tag{5.6}$$

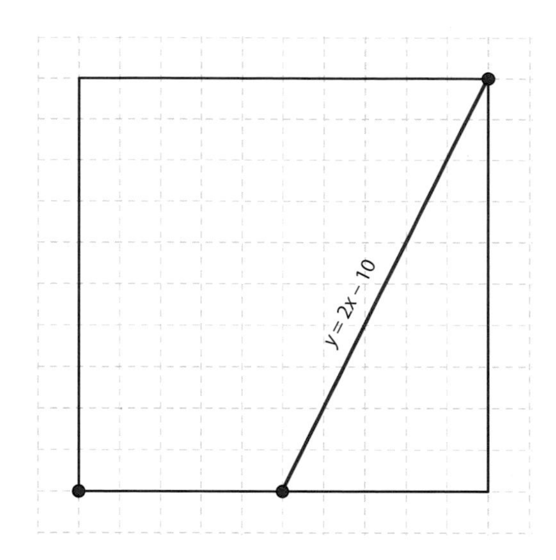

FIGURE 5.15. Hypothetical trajectory.

Finally, this tells us that the intersection point I has coordinates (x, y), where y can be written in terms of x by equation (5.6):

$$x = \frac{(E_1 - D)}{E_2} y + D$$

Substituting this expression for x into equation (5.3) gives us a quadratic in y:

$$\left(\frac{(E_1 - D)}{E_2} y + D - \frac{E_1}{1 - k^2} \right)^2 + \left(y - \frac{E_2}{1 - k^2} \right)^2 = \left(\frac{\| E \|}{k^{-1} - k} \right)^2 . \quad (5.7)$$

It's possible to solve this explicitly, but the answer is a messy expression in terms of E_1, E_2, D, and k. This procedure, however, always gives you the y-coordinate of I, and once you have the y-coordinate, you can get the x-coordinate using equation (5.6).

For example, suppose you spot an enemy missile emerging from the top-right corner, so that $(E_1, E_2) = (10, 10)$. Suppose also that it's headed for the point $(5, 0)$. From this, it follows that the trajectory of the missile is given by the line $y = 2x - 10$. This scenario is illustrated in Figure 5.15.

Let's also suppose that you can fire a missile that's twice as fast as your opponent's missile, so that $k = 1/2$. With this, the equation for S

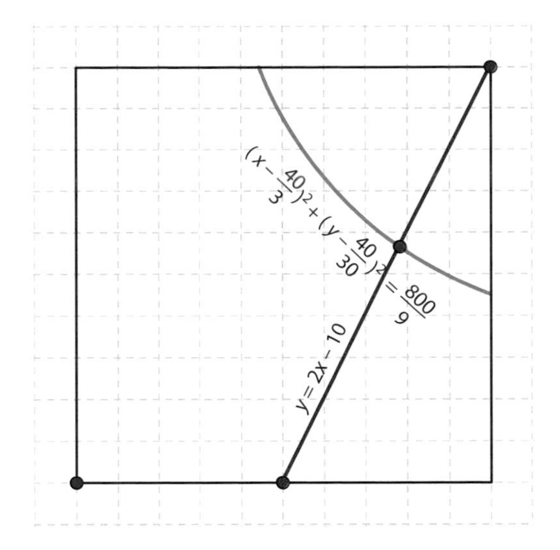

FIGURE 5.16. The coordinates of the intersection are approximately (5.67, 7.84) when $k = 1/2$.

reduces to

$$\left(x - \frac{40}{3}\right)^2 + \left(y - \frac{40}{30}\right)^2 = \frac{800}{9}.$$

In this case, we have a unique intersection point, shown in Figure 5.16. Rather than write down a terrible-looking general expression for y, though, I'll leave you with the following question: since a quadratic equation can have two roots, is it ever possible that you'll be able to intercept your enemy's missile in one of two possible locations? Why or why not?

5.9 BLAST RADIUS

While we've certainly come up with an interesting strategy, it's not actually a strategy for *Missile Command*. In fact, *Missile Command* is much easier than the game we've been studying. The reason is that when you launch a missile in the actual game, it doesn't have to hit its target exactly. Instead, when your missile reaches its destination, it detonates with an explosion that will take out anything within its blast radius. This means we needn't be as precise as our previous analysis would have us believe.

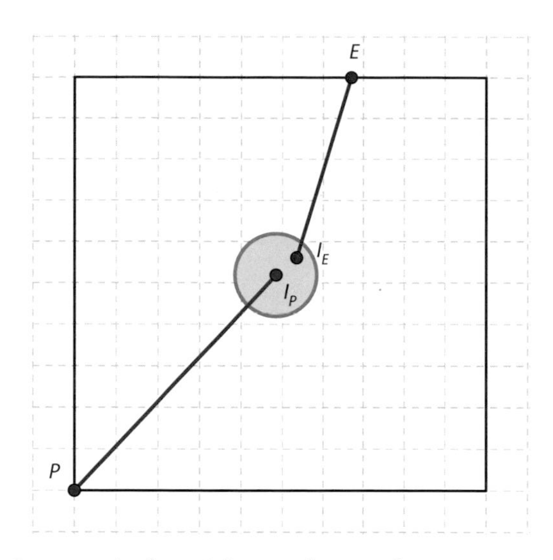

FIGURE 5.17. Our missile doesn't have to be exactly on target—as long as I_E is inside the circle of radius R that's centered at I_P, that missile is going down.

How does this change things? The algebra gets ugly, but the geometry is quite nice. Let's quickly go through the changes.

First, remember how we discovered the Apollonius circle in the first place. It was the representation of the set S of solutions to the equation

$$\frac{IE}{IP} = k,$$

where I represented the point of intersection. Now, however, you may not need to intersect with the missile's trajectory in order to destroy it. All that matters is that the enemy's missile is within the blast radius of your missile (Figure 5.17).

Accordingly, we still let S denote the set of points where it's possible to intercept the enemy's missile. But now the final position of your missile may not be the same as the final position of the enemy's missile. If we let I_P denote the final coordinates of your missile and I_E denote the final coordinates of the enemy's missile, then I_P is in S precisely when

$$\frac{I_E E}{I_P P} = k, \tag{5.8}$$

subject to the restriction that

$$I_E I_P < R,$$

where R is the blast radius. In other words, the coordinates of the enemy's missile should be within the blast radius of your missile. In particular, the set S depends on R—as the blast radius increases, so too does the size of S.

We can get an even clearer picture if we use coordinates. As before, we'll let (x, y) be the coordinates of our missile, i.e., $I_P = (x, y)$. The inequality above means that we can write the coordinates of I_E as

$$I_E = I_P + (\epsilon_1, \ \epsilon_2) = (x + \epsilon_1, y + \epsilon_2),$$

where $\sqrt{\epsilon_1^2 + \epsilon_2^2} < R$. Using coordinate notation and the same argument from before, equation (5.8) then becomes

$$\left(x - \frac{E_1 - \epsilon_1}{1 - k^2}\right)^2 + \left(y - \frac{E_2 - \epsilon_2}{1 - k^2}\right)^2 = \frac{(E_1 - \epsilon_1)^2 + (E_2 - \epsilon_2)^2}{(k^{-1} - k)^2}.$$

Notice that when $\epsilon_1 = \epsilon_2 = 0$, this reduces to equation (5.3), which provides a nice sanity check. Moreover, for fixed values of ϵ_1 and ϵ_2, the above equation still describes a circle. However, since ϵ_1 and ϵ_2 can take on any values subject to the restriction that $\sqrt{\epsilon_1^2 + \epsilon_2^2} < R$, the analogue of S in this case is no longer a single circle but a region in the plane. Figure 5.18 shows how S looks for a fixed blast radius R.

Just as before, we'll be able to intercept the enemy's missile if S intersects its trajectory. Unlike before, however, the intersection no longer consists of a single point—there are now a number of places you can send your missile for a successful hit! Moreover, because S is now a region and has some area,[12] missiles that would have been impossible to intercept in our previous model are now within reach. Figure 5.19 shows some examples with regions S defined by our new model. In each case, the green area represents our new representation of S.

But where should you send your missile once you know your enemy's trajectory? Figure 5.20 shows some sample trajectories for the examples in Figure 5.19. In each case, the lighter green region represents the acceptable target region: if your missile lands anywhere in this region,

FIGURE 5.18. The general shape of our new set S.

you'll get a successful hit. As in the $R = 0$ case, you're welcome to try to describe these sets algebraically (good luck!).

For a game as simple as *Missile Command*, there's some really rich mathematics under the hood. And we haven't even addressed all of it! If you're curious, you may want to think about a few other hidden assumptions we've made. For example:

- We've assumed that you are able to launch your missile as soon as you see your enemy's missile appear on screen, but unless you're a *Missile Command* savant, this isn't likely. Realistically, we'd need to incorporate some reaction time into the model. What effect do you anticipate this might have?
- When a missile detonates (either in real life or on the game), the explosion and subsequent blast radius don't appear instantaneously. Instead, the blast radius grows for some time, reaches a maximum, and then recedes. How might this change the set S? The acceptable target ranges?

5.10 THE PURSUER AND PURSUED IN *MS. PAC-MAN*

For the final stop on our pursuit and evasion tour, let's turn to an even more famous arcade game from the dawn of the modern video game

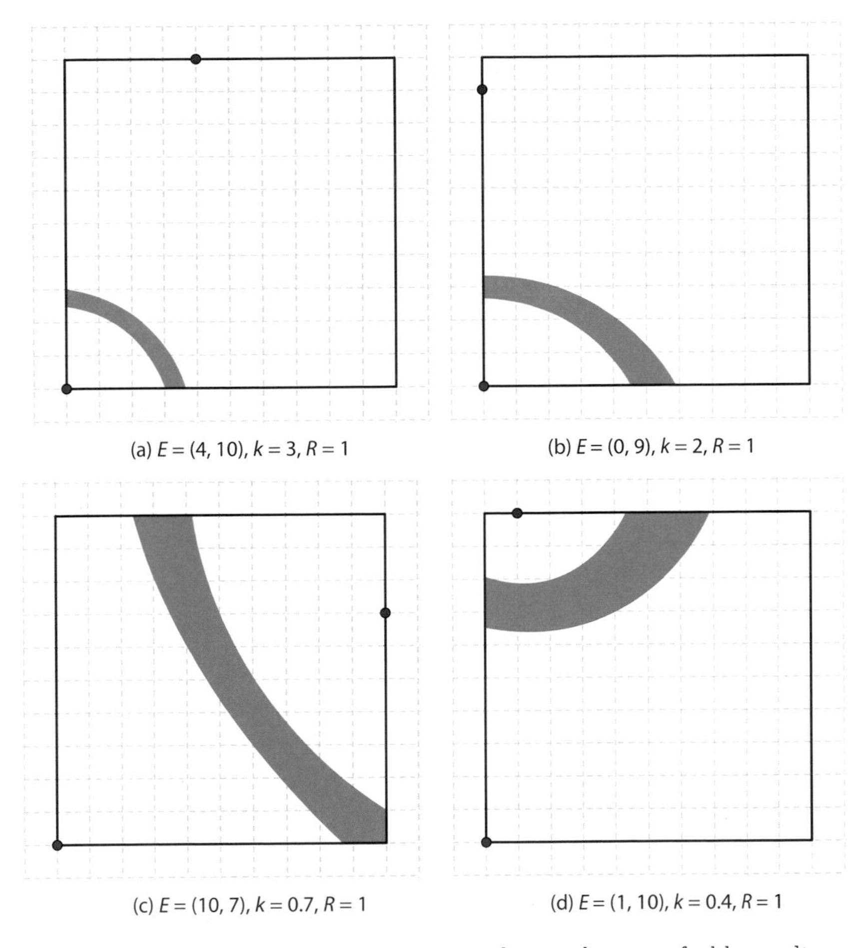

FIGURE 5.19. Modified regions S corresponding to the case of a blast radius.

era. In the history of the medium, perhaps no video game character is more iconic than Pac-Man—and by extension, his more fashionable life partner, Ms. Pac-Man. They have starred in two of the most popular arcade games in history and still have a fair amount of cultural cachet even though they're best known for games released more than thirty years ago.

Both *Pac-Man* and *Ms. Pac-Man* share a similar structure. For a mere twenty-five cents, you take control of one of these characters and navigate through a maze, collecting pellets and avoiding ghosts. Touch the ghosts and you die; collect all the pellets and you advance to the

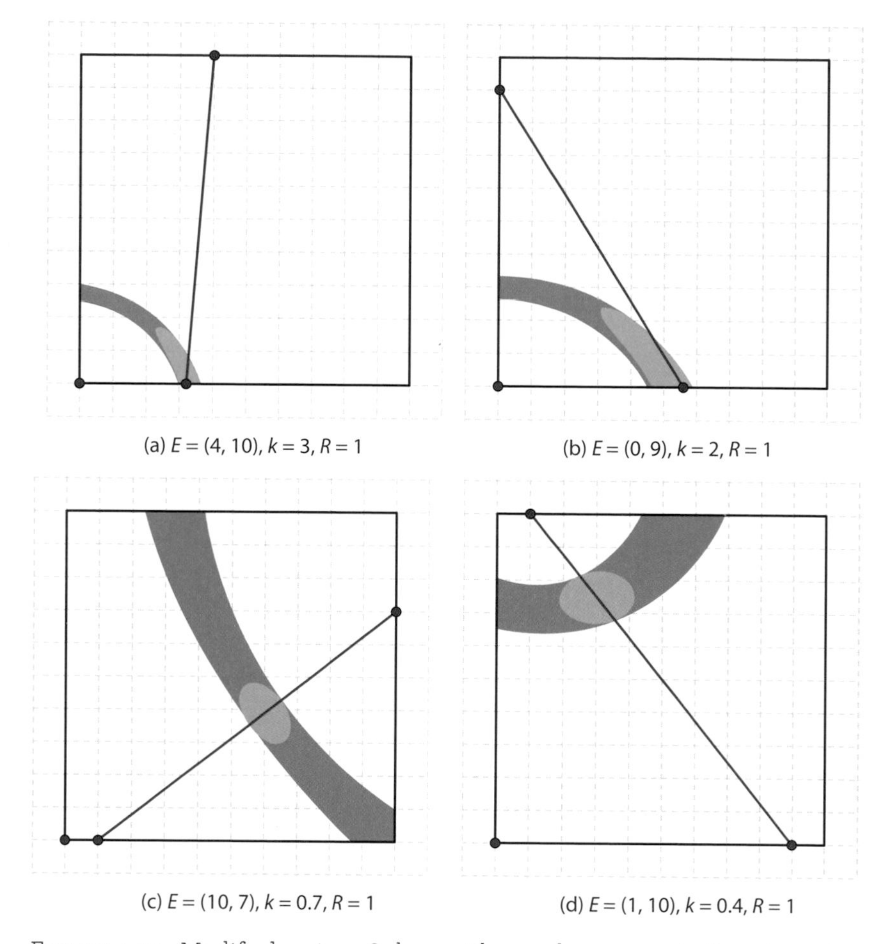

(a) $E = (4, 10), k = 3, R = 1$

(b) $E = (0, 9), k = 2, R = 1$

(c) $E = (10, 7), k = 0.7, R = 1$

(d) $E = (1, 10), k = 0.4, R = 1$

FIGURE 5.20. Modified regions S along with sample enemy missile trajectories.

next level. Moreover, some of the pellets (the aptly named *power pellets*) are larger and turn the tables, allowing Pac-Man to destroy ghosts rather than the other way around.

In other words, these games have a more fluid notion of the pursuer and the pursued. While the player is typically being pursued by the ghosts, he can also become the pursuer for a limited amount of time.

When you're the pursuer, the strategy is clear. If you want a high score, you should eat as many ghosts as possible; if you want to get to the next level, you should eat as many pellets as possible. But what about when you're being pursued? What's the best strategy in this case?

TABLE 5.1. *Ms. Pac-Man* Ghost Behavior

Name	Chase Behavior
Blinky	Blinky targets Ms. Pac-Man's current location.
Pinky	Pinky targets a spot 32 pixels away from Ms. Pac-Man, in the direction of her current motion.
Inky	Take Blinky's location and reflect it across the point that's 16 pixels away from Ms. Pac-Man, in the direction of her motion. Set this location as the target.
Sue	If Ms. Pac-Man is more than 64 pixels away, Sue targets Ms. Pac-Man. Otherwise, Sue targets the lower left corner of the maze.

The answer depends on the behavior of the ghosts. Are they always tracking you, like red shells in *Mario Kart*? Do they move randomly through the maze? Or is something else going on?

The ghost behavior in *Pac-Man* and *Ms. Pac-Man* is similar, but there are some important differences. In *Pac-Man*, the behavior of the ghosts is predetermined, and a knowledgeable player can exploit each ghost's unique pursuit strategy to advance farther in the game.[13] In *Ms. Pac-Man*, the ghosts move randomly during the first seven seconds of play. Afterward, though, they settle into a routine, and each ghost moves according to the rules in Table 5.1.[14]

Needless to say, being able to anticipate the moves of your adversaries is a huge advantage.[15] To see how much of an advantage, researchers at Duke University created an algorithm to play *Ms. Pac-Man* that incorporates the game's ghost behavior. In "A Model-Based Cell Decomposition Approach to On-line Pursuit-Evasion Path Planning and the Video Game Ms. Pac-Man," authors Greg Foderaro, Ashleigh Swingler, and Silvia Ferrari provide a detailed account of their approach to automating gameplay.

There are a number of tradeoffs to consider when playing *Ms. Pac-Man*. For instance, how should a player balance the allure of a power pellet against the risk of moving toward any nearby ghosts? How should a player safely grab a fruit before it disappears, thereby earning some bonus points? When is an attempt on the fruit too risky? Making all these decisions on the fly can be a bit overwhelming for a novice player, but with experience comes a more intuitive grasp of the game's mechanics. Foderaro et al.'s approach works by trying, in some sense, to automate an experienced player's intuition in the face of these decisions.[16]

More specifically, *Ms. Pac-Man* is essentially an exercise in risk management. At every intersection, the player has a choice about which direction to turn. Here are some things that might influence the decision about where to go:

- the number of pellets on each potential path,
- the presence of a power pellet on one of the paths,
- the presence of a fruit on one of the paths (good for bonus points),
- the proximity of each path to neighboring ghosts.

The first three are definite rewards: if one path has a lot of pellets, a power pellet, or a fruit, there's a strong incentive to take that path over one that doesn't have any of these things. But unless the player has recently eaten a power pellet, these three things must be weighed against the last; even if a path is flush with pellets, it won't be appealing if it's also swarming with ghosts.

The algorithm that Foderaro and his colleagues developed takes each of these components into account. Pellets, power pellets, and fruit all contribute to making a path more appealing; proximity to ghosts, however, can easily put a path off-limits. Moreover, because the ghosts' movement patterns are well understood, the algorithm can look several steps out and see how Ms. Pac-Man's choice of direction will influence where the ghosts will head in the future.

In this way, they were able to assign a certain reward value to each possible direction whenever Ms. Pac-Man reaches an intersection, and have her always move in the direction of minimal risk (or maximal reward, depending on your perspective).

Foderaro, Swingler, and Ferrari weren't the first to try to optimize the playing strategy in *Ms. Pac-Man*. However, by incorporating the behavior of the ghosts into their risk function, and by choosing their weights appropriately, over 100 trials their automated playing algorithm achieved a high score of 44,630. In human terms, this isn't so impressive—the world-record high score for *Ms. Pac-Man* exceeds 900,000 points.[17] Among nonsentient players, though, the feat was much more impressive. In fact, at the time, it was a new record. By combining risks and rewards with pursuit and evasion, the authors were able to make Ms. Pac-Man a better informed maze runner.

5.11 CONCLUDING REMARKS

There are plenty of other pursuit-and-evasion examples one could point to, but this sampling should give you a sense for the amount of diversity to be found, both from the mathematics and the video games. Before we move on, one final remark on the *Ms. Pac-Man* paper is in order.

In addition to an interesting example of a more complex pursuit-and-evasion strategy, the paper also provides an example of another rich topic: automated game playing. There are several researchers around the globe who study and design these types of algorithms, but their reasons go beyond simply wanting to break high-score records. Indeed, in many cases the study of these algorithms is tied to one of the most important unsolved problems in modern mathematics. A solution to this problem would bring instant fame and fortune to anyone who provides a solution.

We'll discuss this problem, and its many connections to video games, in the next chapter. Want to learn how to make a million dollars by studying video games? Then read on!

5.12 ADDENDUM: THE PURSUIT CURVE FOR RED SHELLS AND A REFINED INEQUALITY

In this section, we'll derive an equation for the path of a red shell and come up with a more precise condition for when you should wait to fire. This requires a fair bit of calculus, though. The current section is logically independent of the rest of the book, so if things get a little too heady, feel free to skip ahead.[18]

In order to derive an equation for the path of the red shell, we need to capture what it means for the shell to be heat-seeking. One way to interpret this is that the shell's velocity vector is always pointed toward your opponent's kart. This is a fairly natural and intuitive condition—when you're chasing after something, it makes sense to keep your eyes on it and move in the direction of your target. Though red shells in *Mario Kart* games may behave slightly differently, this seems like a reasonable place to start.

Mathematically, this means that the coordinates (x_S, y_S) of your shell must satisfy the following differential equation:

$$\frac{dy_S}{dx_S} = \frac{y_S - y_K}{x_S - x_K},$$

where (x_K, y_K) are the coordinates of your opponent's kart. Note that the right-hand side is equal to the slope of the line between the shell and the kart.

We can say much more than this, however. To start, using our notation and coordinates from before, the position of your opponent's kart at time t is equal to $(0, \ell_K + v_K t)$. Therefore, we can rewrite the equality above as

$$\frac{dy_S}{dx_S} = \frac{1}{x_S}(y_S - \ell_K - v_K t).$$

We know two other things as well. First, after time t, your shell will have traveled a distance $v_S t$. Second, we can compute the distance your shell has traveled using the arc length formula from calculus. In other words, we can express the distance the shell has traveled after time t in two different ways. Equating these two expressions gives us

$$v_S t = \int_0^{x_S} \sqrt{1 + \left(\frac{dy_S}{dz}\right)^2}\, dz.$$

We now have two equations and two unknowns (t and dy_S/dx_S). Isolating t in both equations produces the string of equalities

$$\frac{1}{v_S}\int_0^{x_S} \sqrt{1 + \left(\frac{dy_S}{dz}\right)^2}\, dz = t = \frac{1}{v_K}\left(y_S - \ell_K - x_S\frac{dy_S}{dx_S}\right).$$

Differentiating the left and right sides with respect to x_S gives us

$$\frac{1}{v_S}\sqrt{1 + \left(\frac{dy_S}{dx_S}\right)^2} = \frac{1}{v_K}\left(-x_S\frac{d^2 y_S}{dx_S^2}\right).$$

Next, to simplify the notation, let $f_S = dy_S/dx_S$ and $k = v_K/v_S$. (Note that $k < 1$.[19]) The above equality then becomes

$$k\sqrt{1+f_S^2} = -x_S \frac{df_S}{dx_S}.$$

To solve this differential equation, we separate variables and integrate:

$$\begin{aligned}
\frac{df_S}{\sqrt{1+f_S^2}} &= -k\frac{dx_S}{x_S} \\
\int \frac{df_S}{\sqrt{1+f_S^2}} &= -k\int \frac{dx_S}{x_S} \\
\ln\left(f_S + \sqrt{1+f_S^2}\right) &= -k\ln|x_S| + C_0,
\end{aligned}$$

for some constant C_0, where as usual $\ln(x)$ denotes the natural logarithm function.

We can find C_0 using the initial conditions of the problem. When you launch your shell, $x = -\ell_S$ and $f_S = \ell_K/\ell_S$, so that by properties of the natural logarithm

$$\begin{aligned}
C_0 &= \ln\left(\frac{\ell_K}{\ell_S} + \sqrt{1+\left(\frac{\ell_K}{\ell_S}\right)^2}\right) + k\ln(\ell_S) \\
&= \ln\left(\ell_S^{k-1}\left(\ell_K + \sqrt{\ell_S^2+\ell_K^2}\right)\right).
\end{aligned}$$

Now let's return to our equation

$$\ln\left(f_S + \sqrt{1+f_S^2}\right) = -k\ln|x_S| + C_0.$$

Exponentiating both sides then gives

$$f_S + \sqrt{1+f_S^2} = C\,|x_S|^{-k},$$

where

$$C = e^{C_0} = \ell_S^{k-1}\left(\ell_K + \sqrt{\ell_S^2 + \ell_K^2}\right).$$

It's taken a fair amount of work to get this far, but we're not there yet! Remember, we're trying to derive a formula for the red shell's trajectory, and we currently just have an equation relating dy_S/dx_S to x_S.

To proceed, notice that

$$\frac{C}{|x_S|^k} - \frac{|x_S|^k}{C} = f_S + \sqrt{1+f_S^2} - \frac{1}{f_S + \sqrt{1+f_S^2}} = 2f_S.$$

Moreover, in our coordinate system x_S is always less than or equal to 0, which means that $|x_S| = -x_S$. In other words, we have

$$\frac{dy_S}{dx_S} = f_S = \frac{1}{2}\left(\frac{C}{(-x_S)^k} - \frac{(-x_S)^k}{C}\right).$$

Now we're in the home stretch. Integrating both sides once more, we have

$$y_S = \frac{1}{2}\left(-\frac{C}{1-k}(-x_S)^{1-k} + \frac{1}{C(k+1)}(-x_S)^{k+1}\right) + D$$

for some constant D. Using the initial conditions once again, we know that when $x_S = -\ell_S$, $y_S = 0$, so that

$$D = \frac{1}{2}\left(\frac{C}{1-k}\ell_S^{1-k} - \frac{1}{C(k+1)}\ell_S^{k+1}\right)$$

Putting this all together, we see that the path of the red shell is given by the following equation:

$$\begin{aligned} y_S(x_S) &= \frac{1}{2}\left(\frac{1}{C(k+1)}(-x_S)^{k+1} - \frac{C}{1-k}(-x_S)^{1-k}\right) + D, \\ \text{where } C &= \ell_S^{k-1}\left(\ell_K + \sqrt{\ell_S^2 + \ell_K^2}\right), \\ D &= \frac{1}{2}\left(\frac{C}{1-k}\ell_S^{1-k} - \frac{1}{C(k+1)}\ell_S^{k+1}\right). \end{aligned}$$

(One can substitute the expression for C into the expression for D to obtain some slight simplifications, but we won't bother to do so here.)

While this might look complicated, the y-coordinate of the red shell (y_S) is essentially just a combination of powers of the x-coordinate (x_S). The constants depend on the starting position of each object and how fast the objects travel.

We can now (finally!) use this path to determine a more precise condition under which your shell hits its target instead of the wall. As before, we need the y-coordinate of the shell's path when $x_S = -w/2$ to be less than $w/2$. In other words, your shell will avoid the wall if and only if

$$y_S\left(-\frac{w}{2}\right) < \frac{w}{2}.$$

After some rearrangement of terms, this inequality becomes

$$\frac{2}{w}D < 1 + \frac{1}{2}\left(\frac{C}{1-k}\left(\frac{w}{2}\right)^{-k} - \frac{1}{C(k+1)}\left(\frac{w}{2}\right)^{k}\right). \tag{5.9}$$

In the extreme case $k = 0$, notice that $C - \frac{1}{C} = 2\frac{\ell_K}{\ell_S}$ and $D = \ell_K$, so the equation for y reduces to the linear equation (5.1), and the inequality above reduces to inequality (5.2). In other words, inequality (5.9) can be viewed as a powerful (though perhaps not practical) extension of our earlier work.[20]

6
Gaming Complexity

6.1 FROM RUSSIA WITH FUN

When I was a kid, my mom worried over the content of the games I played. Lest my brain turn to mush, one house rule was that the video games I played should be "brain games." In other words, they should challenge me mentally, rather than appealing to my baser boyhood desires to shoot guns and punch aliens.

Luckily for me, this rule was enforced inconsistently. Thankfully for my mother, though, some great games easily fit into the "brain game" category. One of them was *Tetris*. Released for the Nintendo Entertainment System in 1989, *Tetris* has become one of the most well known games of all time, having sold hundreds of millions of copies across different platforms in the years since its initial release.

If you've never played *Tetris*, that's okay! This just means you have one of life's great pleasures awaiting you. I'd encourage you to take a few minutes and try the game out for yourself. Free versions of *Tetris* are just a Google search away.

In the game, players manipulate shapes that mathematicians would call *tetrominoes*. These are shapes composed of four linked squares. There are seven ways the squares can be arranged, yielding the seven well-known *Tetris* shapes: the O, the I, the T, the L, the J, the S, and the Z the (Figure 6.1).[1]

Tetris is played in a rectangular grid. The tetrominoes fall into this grid, one piece at a time. The goal is to arrange the tetrominoes so that they fit nicely in the game board. Whenever the player fills a row of the board with squares, the row is cleared and the player receives points. If the player fills multiple rows at the same time, even more points are awarded, up to a maximum of four rows. Clearing four rows at once is called a *Tetris* line clear (Figure 6.2). The more rows you clear before

FIGURE 6.1. Can you tell which shape corresponds to which letter?

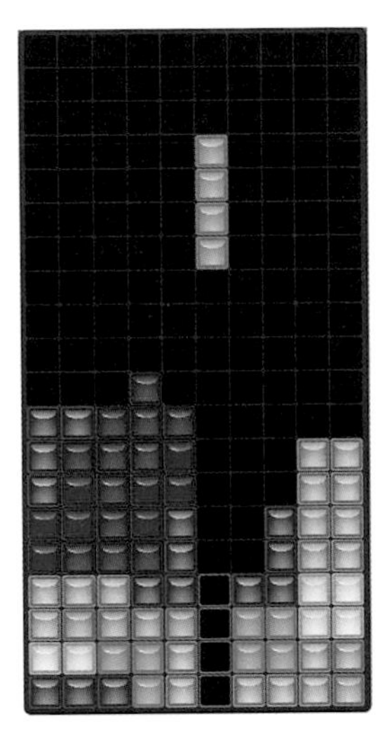

FIGURE 6.2. A Tetris line clear will occur when the light blue block lands.

the board fills up, the more points you score. What makes the game challenging is how the game responds to your success: the more rows you clear, the faster the pieces begin to fall.

Descriptions of *Tetris* are much less compelling than the experience of playing it yourself. Beneath the simple mechanics lies a rich gameplay experience. Play too much and you may begin to see the world in terms of stacking geometric shapes. This phenomenon is common enough that it's even been given a name: *Tetris* syndrome. So strong is the influence of the game on our thoughts that some studies have suggested playing *Tetris* can help reduce the severity of PTSD symptoms.[2]

Despite a lack of sci-fi weaponry or loud explosions, *Tetris* has more than enough to engage its audience. Because it requires both a quick mind and quick reflexes, it was a hit with both my mother and me.

The game is not without its frustrations, though. When the shapes start falling quickly, it's easy to make a costly mistake. While I certainly played a fair bit of *Tetris* when I was younger, I can't say that I enjoyed every minute of it.

Modern-day *Tetris* games have additional features that try to make the game less frustrating. For example, some versions of the game now allow you to "hold" one of the shapes in a staging area, where you can pull it out and use it whenever you want. This is particularly useful if you want to stockpile a coveted I piece for later use to clear four rows at once. And while all versions of the game show you which piece is coming next, some versions show the next several pieces, so that you can plan a bit better.

Some may argue that these compromises make the game easier, but in absolute terms, the game is hard no matter what. In fact, even if a version of *Tetris* came out that showed you every block to come, not just two or three, the game would still be a challenge. There's a mathematical truth lurking in here, and it's one that makes an appearance with surprising frequency in video games if you know to look for it.

6.2 P, NP, AND KEVIN BACON

Describing the mathematics hidden in *Tetris* is certainly possible, but it's not necessarily the most accessible entry point for what's coming. Instead, let's take a step back and consider a (seemingly) different problem from the world of film.

Kevin Bacon is a prolific actor, with a career that's spanned five decades and dozens of films. He's also the focal point of the party game Six Degrees of Kevin Bacon. The goal of this game is to connect an actor to Kevin Bacon through a chain of intermediate actors, where each actor is connected to the previous one via a film they worked on together. The smallest chain wins, and the number of films in that chain is defined to be the *Bacon number* for the other actor.

The game has spawned a number of variants. One popular alternative is the *Wikipedia Game*, an online game in which players try to connect two seemingly disjointed Wikipedia articles by clicking through intermediate pages. For instance, players might have to connect *"Chinese zodiac"* to *"Dallas, Texas"*; the player who can connect these two through the smallest number of links within Wikipedia wins the round.

Six Degrees of Kevin Bacon works in the same way, but with actors rather than Wikipedia articles. For example, suppose we want to connect Kevin Bacon to fellow A-lister Samuel L. Jackson. It's possible to connect the two through the following chain: Jackson was in *Pulp Fiction* (1994) with John Travolta, who starred in *Hairspray* (2007) with Queen Latifah, who in turn was in *Beauty Shop* (2005) with Kevin Bacon. In other words, Samuel L. Jackson's Bacon number is at most three.

But we can do better than this. In fact, Jackson's Bacon number is actually two. One way to see this is to go through Mark Ruffalo, who starred with Jackson in *The Avengers* (2012), and with Bacon in *In the Cut* (2003).

Of course, I haven't done anything to prove to you that Jackson's Bacon number is two; all I've done is show you that his Bacon number is at most two. But if you don't believe me, all you need to do is check that Jackson has never been in a film with Kevin Bacon. This would mean that Jackson's Bacon number must be greater than one: since we already know it can't be greater than two, this would prove that it equals two.

Checking Samuel L. Jackson's name against Kevin Bacon's body of work may seem like a tedious task, but it's something that modern computers can do with ease.[3] Moreover, this method can be used to find the Bacon number to any actor:

1. Pick an actor, *A*, whose Bacon number you want to find.
2. If *A* has been in a movie with Kevin Bacon, *A*'s Bacon number is 1 (by definition, only Kevin Bacon has a Bacon number of 0). If not, create a list of all actors who *have* been in a film with Kevin Bacon.
3. If *A* has been in a movie with someone on this new list, then *A*'s Bacon number is 2. If not, create a list of all actors who have been with someone on your list from step 2.
4. Repeat this process until *A* has been in a movie with someone on your list. The number of connections (i.e., movies) from *A* to Kevin Bacon is *A*'s Bacon number.

Again, this may sound like a tedious procedure, and indeed it would be if you had to do it all by hand. But for a computer, all of this searching can be done in fractions of a second. In other words, in the grand scheme of things, finding someone's Bacon number is not so hard.

But now, let's turn things around and ask a slightly different question. The Bacon number represents the shortest path between an actor and Kevin Bacon within the network of movies. But what if we want to find the *longest* path between an actor and Kevin Bacon? To put it another way, for a given actor (e.g., Samuel L. Jackson), we can formulate a general yes-or-no question as follows: does there exist a path to Kevin Bacon that's of length k or more for some large value of k? The smallest value of k that works is the actor's Bacon number, but now we want to make k as large as possible.

Some ground rules are in order before we consider this new rule. The example above might suggest that there are paths of infinite length. After all, if we wanted to take as long as possible to connect Samuel L. Jackson to Kevin Bacon, couldn't we just bounce back and forth between John Travolta and Queen Latifah as many times as we wanted? To avoid this loophole, we allow our paths to go through an actor only one time (to be mathematically precise, we say the paths must be *simple*). Note this also means that we can't hit the same actor twice even if we go through different movies—for example, even though John Travolta and Samuel L. Jackson have been in *Pulp Fiction* (1994) and *Basic* (2003) together, once we've made a connection between the pair through one of these movies, we can't later use the other movie to build a second connection.

This problem—let's call it the Max Bacon problem, for short—is much trickier to solve. Is there a path of length 100 or more (i.e., a sequence of connections through distinct actors that passes through 100 movies) from Samuel L. Jackson to Kevin Bacon? What about a path of length 1,000 or more? 10,000 or more? Unlike calculating the Bacon number, there's no clear strategy here that can find these paths in a reasonable amount of time.

There are *brute force* approaches that would work in principle. For example, we could find all paths of length $k - 1$ and see whether Jackson and Bacon can bookend any of them to form a path of length k. But with so many actors in the world (past and present), coming up with these lists isn't feasible even for moderately sized k (say, $k \approx 1{,}000$).[4]

TABLE 6.1. A Long Path from Samuel L. Jackson to Kevin Bacon

Starting Actor		Ending Actor
Samuel L. Jackson	→	Jamie Foxx
Jamie Foxx	→	Emma Stone
Emma Stone	→	Viola Davis
Viola Davis	→	Craig Robinson
Craig Robinson	→	Ben Stiller
Ben Stiller	→	Janeane Garofalo
Janeane Garofalo	→	Zooey Deschanel
Zooey Deschanel	→	Joseph Gordon-Levitt
Joseph Gordon-Levitt	→	Ellen Page
Ellen Page	→	Kevin Bacon

It's not all bad, though. While it may seem difficult to find a solution, it's not difficult to check the *validity* of a solution. For example, suppose I tell you that there's a path of length ten from Samuel L. Jackson to Kevin Bacon through the actors in Table 6.1.

It's easy to check (or have a computer check) that this path is valid. With some Internet sleuthing, you could probably verify this chain in a few minutes.[5] A computer could verify it even faster.

So even though *finding* a path of length k or greater seems difficult, *confirming the existence* of a path of length k or greater is easy. If you're a lucky guesser, maybe you can find a really long path between Samuel L. Jackson and Kevin Bacon without much difficulty. But hoping for a lucky guess won't make for a good algorithm.

The difference between finding an actor's Bacon number and solving the Max Bacon problem lies at the heart of one of the most famous unsolved problems in computer science. Finding an actor's Bacon number—that is, finding the shortest path between an actor and Kevin Bacon—is essentially an example of the *shortest path problem*, which is concerned with finding the shortest path between two nodes in a graph. This problem belongs to a class of problems bestowed with the minimally descriptive moniker **P**. Very loosely, problems in **P** are ones that computers can solve relatively quickly.[6] On the other hand, the Max Bacon problem is an example of the *longest path problem*. As you might have guessed, this problem is concerned with finding the longest path between two nodes in a graph. Unlike the shortest path problem, the longest path problem doesn't live in **P**. Instead, the longest path problem belongs in the problem class **NP**. Informally, problems in **NP** are ones whose solutions we can verify quickly.[7]

It's clear that every problem in **P** is also in **NP**. After all, if we can solve a problem quickly, the process of solving the problem also verifies the validity of the solution. But the opposite direction is profoundly less obvious: is every problem in **NP** also contained in **P**? Or, in simpler (though less precise) terms, if we can verify the solution to a problem quickly, does this mean we can also *solve* the problem quickly?

If **P** = **NP** (that is, if these two sets of problems are really the same), it would have profound implications. It could lead to algorithms that would bring about tremendous advances in biology, economics, logistics, and even mathematics itself. For instance, since many of the current cryptographic systems around the world are based on the assumption that there are problems in **NP** but not in **P**, much of the infrastructure in place for things like online banking and shopping would have to change dramatically.

The implications are so profound, in fact, that many people believe that they're too good to be true. In other words, many people who have an opinion on the matter believe that **P** ≠ **NP**.[8] But belief, no matter how strong, does not a proof make. And even though mathematicians and computer scientists have been studying the **P** vs. **NP** problem for decades, a proof one way or the other has remained tantalizingly out of reach. In fact, it is one of seven Millennium Prize Problems listed by the Clay Mathematics Institute in 2000. Solution of any one of these problems, **P** vs. **NP** included, comes with instant fame and worldwide recognition (at least within certain circles). It also comes with a $1 million prize. There are many excellent resources on this problem if you're looking for more than the broad overview provided here.[9]

One final remark, before we proceed: within the class of **NP** problems, there is a distinguished subclass of problems, known as **NP-complete**. These problems are so named because any **NP** problem can be transformed into an **NP-complete** problem. This means that if we can show that a single **NP-complete** problem is in **P**—that is, if we can solve the problem quickly, not just verify a solution quickly—this means that **P** = **NP**, since any problem in **NP** could first be transformed to the

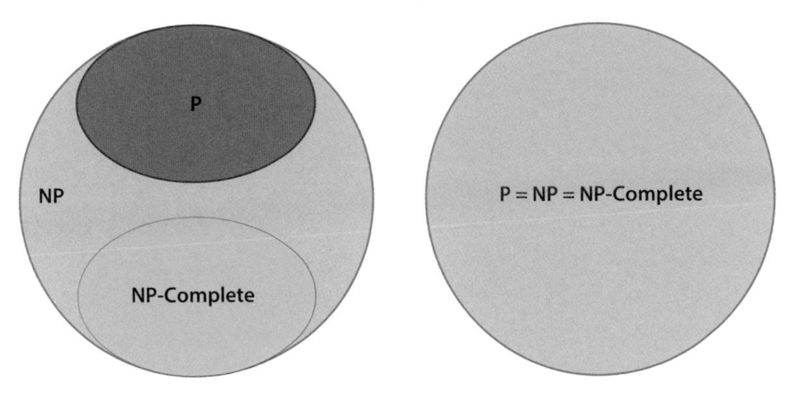

FIGURE 6.3. One of the scenarios above represents the truth. The left represents the relationships between **P**, **NP**, and **NP-complete** problems if **P** $\neq$ **NP**. The right represents the relationships if **P = NP**. The fact there are **NP** problems which are not **NP-complete** if **P** $\neq$ **NP** was proven by Richard Ladner in 1975.

solved **NP-complete** problem (Figure 6.3). In other words, just a single algorithm could be the key to proving that **P = NP**!

6.3 DESKTOP DIVERSIONS

But what does any of this have to do with *Tetris*? As it turns out, *Tetris* is hard. Not just in a controller-tossing, tantrum-inducing sort of way, but in a way that's mathematically precise.

In 2003, researchers at MIT published an article titled "Tetris is Hard, Even to Approximate" (Breukelaar et al. 2014). The article showed that many problems one typically wants to solve when playing *Tetris* are in **NP**. More precisely, they showed that even if you had *complete knowledge* of the sequence of pieces that would show up (rather than just knowing the next piece, or the next several), the following problems are all **NP-complete**:

- clearing as many rows as possible;
- playing as many pieces as possible before the game ends;
- creating as many *Tetris* line clears as possible (i.e., moves that clear four rows simultaneously);
- keeping the pieces as low as possible, to avoid the top of the grid and the inevitable end of the game.

To summarize what we already know, the fact that these problems are **NP-complete** tells us a few things. First, it means that we don't have an algorithm that can "quickly" tell us how to play *Tetris* optimally. Of course, if $\mathbf{P} \neq \mathbf{NP}$, then no such algorithm even exists. On the plus side, if we think we have a really great sequence of moves (say, one that will result in at least 50 *Tetris* line clears), this result tells us that we can verify our solution relatively quickly.

The fact that these problems are **NP-complete** and not just **NP** is also significant. In theory, this means that if we find an algorithm to quickly solve one of these problems, we will have proven that **P = NP**! So in some sense, playing a whole lot of *Tetris* may be the first step toward solving one of the most famous problems in mathematics—to say nothing of becoming a millionaire. Before settling in for a marathon session, however, remember that there are risks involved: if $\mathbf{P} \neq \mathbf{NP}$, no amount of playing will help you find an algorithm that works (though it may give you some ideas as to why such an algorithm can't exist).

If *Tetris* isn't really your thing, fear not. There are plenty more **NP** problems that arise in games. For example, another game that we've collectively spent millions of hours playing is *Minesweeper*. Included in every version of Microsoft Windows for 20 years—from version 3.1 in 1992 until Windows 8 in 2012—the game puts you in the otherwise unenviable position of trying to find a certain number of land mines inside of a rectangular grid. Figure 6.4 shows an example of one such grid.

On the left of Figure 6.4 is an unmarked board. This is how the game begins: we're presented with a grid of tiles, any one of which we can select. Once we select a tile, the game begins.

The grid on the right represents an example of the board after one tile has been selected. Tiles with numbers inside tell us how many adjacent tiles have mines. Blank tiles (such as the one in the bottom right corner) tell us that there are no adjacent mines with respect to that tile.

From the arrangement of tiles, we can deduce the location of many of the mines. For example, the tile highlighted in red must have a mine beneath it, because the tile above has a value of 1, and none of its adjacent tiles are covered. Using similar reasoning, we can mark a few more tiles as ones that definitely have mines beneath them. By

FIGURE 6.4. On the left: a blank *Minesweeper* board. On the right: one example of the board after a single click. Screenshots from *Minesweeper*. © Microsoft

FIGURE 6.5. What's wrong with this picture? Screenshots from *Minesweeper*. © Microsoft

repeating this process of identifying mines and non-mines, we can eventually complete the puzzle.

Now, imagine for a moment you were playing a game of *Minesweeper* and found yourself with the board depicted in Figure 6.5. In the figure, there are four mines and four hidden squares. Clearly the mines must be behind the squares. However, the values around the squares aren't correct—specifically, the 2s on the corners should be 1s. If you were ever

FIGURE 6.6. Is anything wrong with this picture? Based on Microsoft's *Minesweeper*.

presented with an image like this while playing *Minesweeper*, you could rightly conclude that there's something wrong with the game.

The mistake in the example above was easy to spot. But what if you're playing a more advanced board? Take the one in Figure 6.6, for instance.

Presumably any game of *Minesweeper* you're playing won't be buggy enough that an inconsistent board—one in which the numbers don't match the placement of the mines—would appear. But what if you wanted to be *extra* sure? What if you wanted to develop an algorithm that could look at a board and tell you whether or not it's logically consistent?

This may seem like not a particularly useful algorithm, but in fact, with such an algorithm you could solve *Minesweeper* boards much more quickly. Any time you wanted to place a flag over a tile to indicate a mine, you could run the algorithm afterward; if it tells you that the board is inconsistent, that means you must have made a mistake placing the flag. In fact, such an algorithm could be used to solve many *Minesweeper* boards.[10]

Sounds great, right? Here's the problem: being able to identify a *Minesweeper* board as consistent or not is **NP-complete**. Mathematician Richard Kaye proved this in 2000. In other words, we have no idea how to create such an algorithm, or even if one exists. If you really love *Minesweeper*, though, perhaps it's something worth thinking about.

6.4 PLATFORMING PROBLEMS

Tetris and *Minesweeper* aren't the only games that contain **NP** problems. In fact, many classic games from the 8- and 16-bit generations have ties to **P** vs. **NP**.

Nowadays it's fairly easy to find and play classic games from the 1980s and 1990s. Nintendo's Wii and Wii U both come with virtual consoles that host a fairly thorough backlog of the company's classic games, from series like *Mario*, *Zelda*, *Donkey Kong*, and *Metroid*. Nintendo even released a miniature version of its original NES console in 2016, preloaded with thirty classics. For those of you looking to bone up on your video game history, you couldn't do much better than playing some of these games.

Three of these series are typically classified as platformers. Much of the action in games like *Super Mario World*, *Donkey Kong Country*, and *Metroid* involve running, jumping, dodging, and attacking enemies while making your way from point A to point B. By contrast, games in the classic *Zelda* series use a top-down perspective and more open-world exploration (though this latter feature is also shared by *Metroid* games).

Many people play these games and, after finishing them, are content to move on to new ones. Ardent fans may play a game several times to uncover all of its secrets. For a select few, though, mastery isn't enough. In fact, there's an entire subculture of people who have hacked into these classic games and created new levels within them. The particularly sadistic ones create levels of such maddening complexity that, were it not for YouTube videos proving otherwise, they would seem impossible to complete.[11]

In fact, major developers are beginning to bet on the growth of this budding level-design community. In 2015, for example, Nintendo released a game called *Super Mario Maker* for Wii U. The game allows players to build their own levels using assets from the *Mario* series, and play those levels with friends. During a press event after the game was announced, Nintendo showcased some particularly difficult levels.[12] And indeed, the launch trailer for the game previewed some levels that seem simply impossible.

Suppose you found yourself playing through a particularly difficult level created in *Super Mario Maker* (or one of its unlicensed brethren). After dying a few times, you might rightfully conclude that the level is difficult. Several hours later, with countless deaths under your belt, you may start to wonder whether or not victory is even possible. But no matter how many times you die, repeated failure may not be enough to *prove* that a level is impossible—maybe you're just not as good at the game as you need to be.

In a situation like this, it would help to know whether or not the level is *actually* impossible within the physical reality of the game world. Put another way, it would be nice if there were algorithms to tell us whether it's possible to complete a level, given its design and the rules for how the game character moves. Unfortunately, here again we find that **NP** rears its ugly head. As proven by a group of researchers in a paper titled "Classic Nintendo Games Are (Computationally) Hard," determining whether it's possible to complete a level designed in any of these games is an **NP** hard problem.[13] In fact, for *Super Mario World* and games in the *Donkey Kong Country* series, deciding whether or not a level is impossible is **NP-complete**. While it's possible to finish all the levels created for the original games, determining whether or not a custom-built level is impossible or not is a very difficult problem, in general.

6.5 FETCH QUESTS: AN OVERVIEW

Let's look at one more example, this time using a game mechanic known as the *fetch quest*.

In many games, collecting things is part of the gameplay. Sometimes there are rewards for collecting items that enhance the experience, while other times collecting is required just to advance the story. Still other times, they serve no real purpose except to create an itch to scratch for the "completionist," someone who can't stop playing a game until every facet of it has been thoroughly explored.

FIGURE 6.7. Screenshots from *inFAMOUS 2*. The character's maximum health is represented by the jagged blue curve. Early in the game (left), this curve is short, but as the character grows stronger, it lengthens. © Sony

Examples of fetch quests abound: almost every best-selling console game of the past ten years has featured one in some form or another. Here are a few quick examples, each highlighting one of the fetch-quest categories described above.

One series that ties fetch quests to in-game rewards is *inFAMOUS*, which has spawned three games across two Sony platforms (the PlayStation 3 and PlayStation 4). In these games, the player takes on the role of someone with superpowers, and players get to decide whether to use those powers for good or evil.

Regardless of the decision, the player spends a fair amount of time fighting off enemies. As with most origin stories, initially the player's character is relatively weak, but the character becomes more powerful over the course of the game. In the first two games, one way to grow stronger is by increasing the amount of damage the character can take before dying.

In order to increase the character's health, the player needs to collect items called *blast shards*.[14] You'll get some of these shards by playing through the main story, but the vast majority of them are scattered throughout the game's open world. Finding them can be difficult and time-consuming and is entirely optional. However, collecting them yields a noticeable enhancement to the character: a longer health bar (Figure 6.7). So even though the story doesn't provide much reason to search for these shards, the gameplay does.

On the other hand, one of the better known examples of a fetch quest with some narrative weight behind it comes from a game in the *Zelda* series. In 2003, Nintendo released *The Legend of Zelda: The Wind*

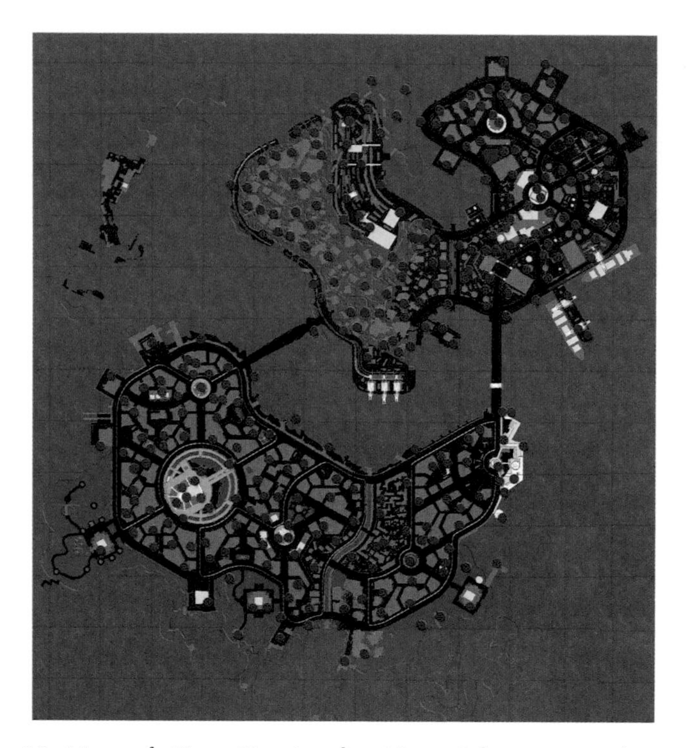

FIGURE 6.8. Map of New Marais, the New Orleans-inspired setting of *inFAMOUS 2*. Blast shards are indicated on the map as red dots. As evidenced by the image, collecting them all can be kind of a slog. © Sony

Waker for the GameCube. It received critical acclaim in spite of an art style that initially divided its fans. The game has withstood the test of time, and ten years later Nintendo rereleased the game in HD for Wii U.

One of the criticisms levied at the game involved a fetch quest. Late in the story, the hero named Link needs to find eight pieces of an ancient artifact called the Triforce. To find these pieces, Link first needs to find treasure maps to each piece's location. And if that weren't enough, the treasure chests can't be deciphered without the aid of an extortionist who charges Link a large amount of in-game currency to translate the maps into something Link can use.

In other words, this portion of Link's quest consists of the following steps: (1) find eight unintelligible maps scattered throughout a large game world; (2) earn enough money to pay for each one to be deciphered; (3) go to eight *new* locations in the game world to acquire the Triforce pieces.

If this sequence of events sounds tedious, that's because it is. There's no compelling narrative force driving this sequence of steps, and while this fetch quest extends the length of the game, it does so in a way that feels like a cop-out. Even Nintendo acknowledged the monotony, and in the HD remake, five of the eight treasure maps were eliminated. Instead, Link could acquire a majority of the Triforce pieces by finding them directly, rather than jumping through the aforementioned hoops.[15]

Another example of a story-driven fetch quest can be found in the *Grand Theft Auto* series. Developed by Rockstar Games, this series is known for its open-world, sandbox style of gameplay. In each game, the player is thrown into a modern-day metropolis, heavily inspired by actual cities like New York and Los Angeles. Players are free to explore the city, or wreak havoc by stealing cars, punching people, and doing all manner of things that one could never get away with in real life.

To help immerse the player in these open worlds, more recent games in the series populate their cities with plenty of hidden collectibles. Consider, for example, *Grand Theft Auto V*, which was released in 2013. In this game there are many different types of collectibles. One is a set of fifty letter scraps strewn throughout the game's Southern California-inspired locale of Los Santos (Figure 6.9). While not central to the main story line, the scraps revolve around a murder mystery; collecting them all solves the mystery and opens up an additional mission.

If solving murder mysteries sounds like your idea of a good time, you might also be interested in Rockstar Games' *L.A. Noire*, released in 2011. In *L.A. Noire*, players take on the role of a 1940s officer in the Los Angeles Police Department. While the game shares the exploration-encouraging open environments of the *Grand Theft Auto* series, it also discourages (or in some cases, prohibits) players from any sort of behavior that an upstanding member of the Los Angeles Police Department wouldn't engage in.

This game provides an example of a fetch quest without any narrative or gameplay payoffs. The quest involves golden film reels; there are fifty hidden throughout the city. Collecting them all provides no gameplay or narrative benefit, but they're there nevertheless for the compulsive collector.

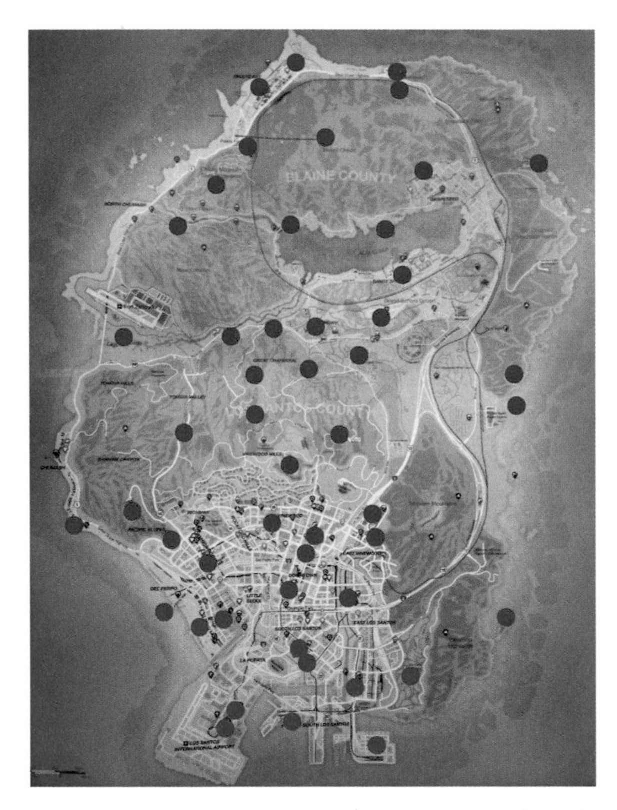

FIGURE 6.9. A map of Los Santos in *Grand Theft Auto V*. The red dots indicate letter scrap locations. *Grand Theft Auto V* & *L.A. Noire* screenshots, courtesy of Rockstar Games, Inc.

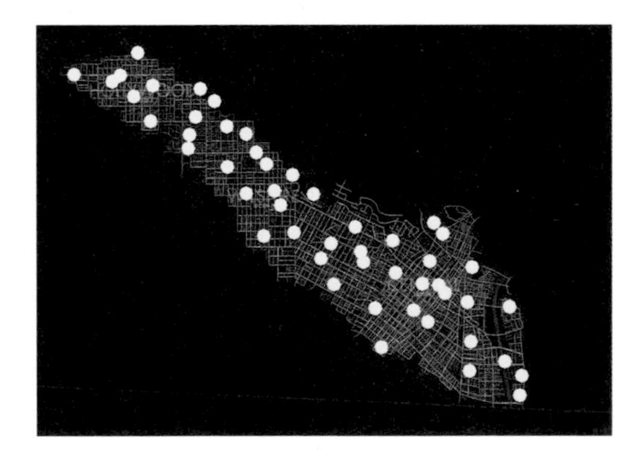

FIGURE 6.10. A map of Los Angeles in *L.A. Noire*. The yellow dots indicate golden film reel locations. *Grand Theft Auto V* & *L.A. Noire* screenshots, courtesy of Rockstar Games, Inc.

6.6 FETCH QUESTS AND TRAVELING SALESMEN

As you can see, fetch quests can be fairly mundane affairs, and the payoff (when there is one) for completing such a quest may not be worth the investment in time.

Nevertheless, for some, the completion of every fetch quest in a game is part of mastering it. For example, in Chapter 4 we briefly discussed speedrunners, for whom getting just a little bit faster in a game can make the difference between a world record and a forgettable play session. If you're a speedrunner, there's value in being able to complete a fetch quest as quickly as possible. With careful planning, maybe one could map out the fastest route through a fetch quest in order to minimize the time spent doing it.

Regardless of the motivation (or lack thereof) behind a fetch quest, there's still the question of what it has to do with the **P** vs. **NP** problem. To make this connection, and to highlight how prevalent these quests are in games, let's look at yet another example, this one from the *Assassin's Creed* series. We've discussed this series already, but what I neglected to tell you is that the games in this series feature a litany of fetch quests.

The quest we'll consider here is from *Assassin's Creed: Revelations*, a 2011 release that takes place in sixteenth-century Constantinople. You take on the role of middle-aged assassin Ezio Auditore, an Italian who comes to Constantinople to do a little family history research, and also to kill tons of people. Littered throughout this world are one hundred collectible "data fragments," which—well, really, does it matter? There are one hundred things to collect, and you can choose to collect them or not.

The game opens in the Galata area of the city, which is separated by sea from the rest of Constantinople. It's a relatively small portion of the game area, but it contains eleven of its data fragments. Figure 6.11 shows a map of the area: the eleven fragments are indicated by orange dots, while the green dot represents the assassin's headquarters in Constantinople. This is Ezio's home base.

If you're trying to collect all of these data fragments, what's the best way to go about it? To put it another way, suppose that Ezio leaves the assassin's headquarters one day with the goal of collecting all of these

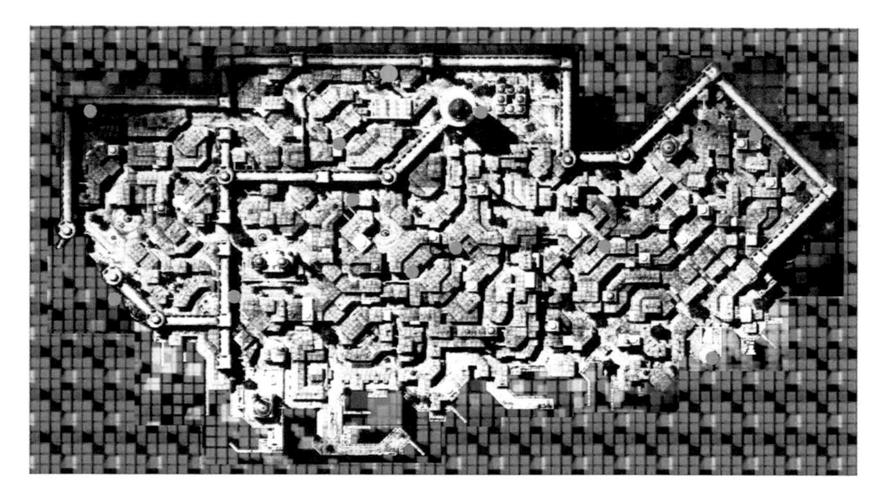

FIGURE 6.11. The Galata neighborhood in Constantinople (present-day Istanbul). Screenshot from *Assassin's Creed Series*. © Ubisoft Entertainment. All Rights Reserved. *Assassin's Creed*, Ubisoft, and the Ubisoft logo are trademarks of Ubisoft Entertainment in the U.S. and/or other countries.

fragments—what's the fastest route he can take to gather the fragments and return home?

As it turns out, this is an example of a very famous mathematical problem known as the *traveling salesman problem* (or TSP, for short). The more familiar context involves, unsurprisingly, a traveling salesman, whose goal is to visit a given list of cities using the shortest path possible (to minimize costs). More precisely, given a list of cities, the problem asks what's the shortest path that visits each city exactly once, and returns to the home city?

This is a hard problem, and finding a general algorithm to solve it would have profound consequences. In fact, not only is the problem hard, it's **NP-hard**, which (informally) means that it's at least as hard as the hardest problems in **NP**. There is, however, a related problem that is **NP-complete**: namely, for a given length L, asking whether or not there's a path the salesman can take that is shorter than L. That is to say, the traveling salesman problem is intimately related to the other examples we've considered here. Intuitively, it should make some sense that this problem would be in **NP**: if someone gives you a path, it's easy to check whether or not its length is less than L. However, if the number of cities is large, actually finding a path of length less than L seems like it could be quite difficult.

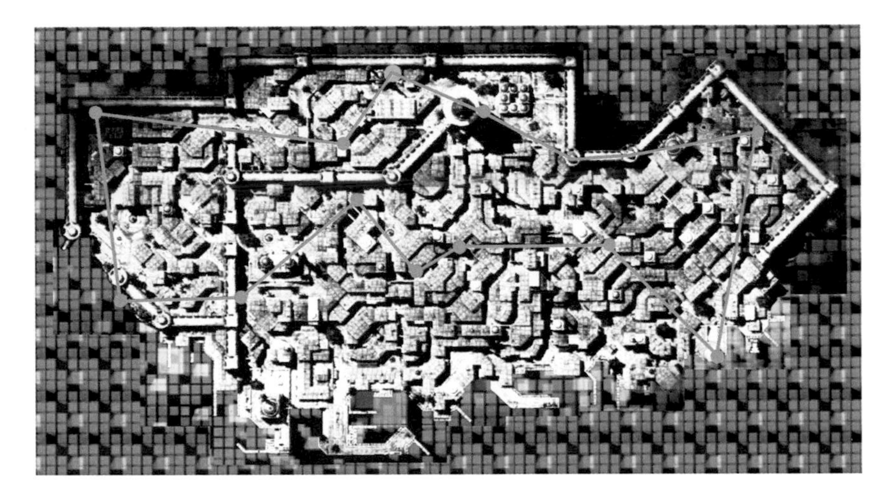

FIGURE 6.12. Shortest path to the eleven data fragments, starting and ending with the assassins' headquarters. Screenshot from *Assassin's Creed Series.* © Ubisoft Entertainment. All Rights Reserved. *Assassin's Creed,* Ubisoft, and the Ubisoft logo are trademarks of Ubisoft Entertainment in the U.S. and/or other countries.

In *Assassin's Creed: Revelations,* the minimal distance Ezio would need to travel is an example of the traveling salesman problem. In Galata, since there are eleven data fragments, we have eleven choices for the fragment to collect first, ten choices for the fragment to collect second, and so on. This means the total number of possible paths is equal to

$$11 \times 10 \times 9 \times 8 \times 7 \times 6 \times 5 \times 4 \times 3 \times 2 \times 1 = 11! = 39{,}916{,}800.$$

Finding the shortest path among those nearly 40 million options isn't so hard for a computer to do, but this number grows more than exponentially as the number of cities grows, as we will see in just a moment.

For the present example, however, it's not hard for a machine to find the shortest path. In fact, the path is shown in Figure 6.12.[16]

Of course, this path is only a two-dimensional representation of what's actually a three-dimensional route. Some of the data fragments may be high on top of buildings, while others are lower to the ground. However, incorporating the extra dimension would add unnecessary complexity; this is a pretty good first pass for a road map to follow.

Or is it? In fact, we've overlooked a couple of things that could help make the path even shorter (although in general, they don't bring us

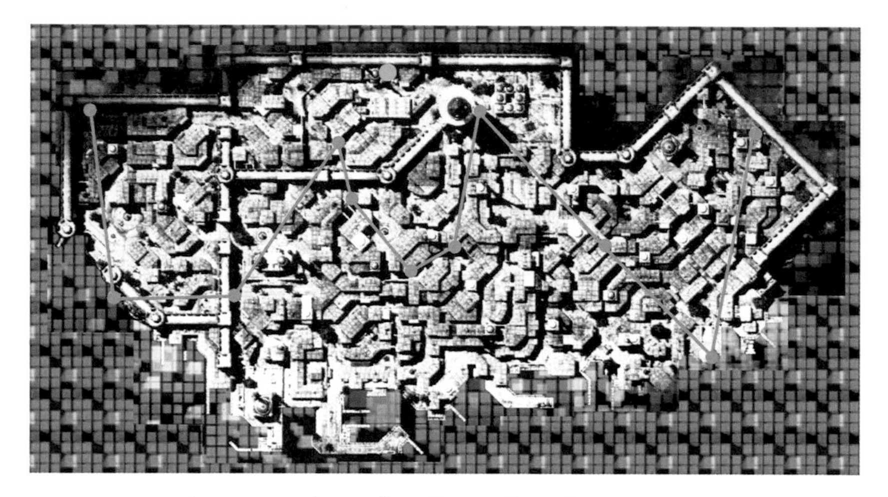

FIGURE 6.13. Shortest path to the eleven data fragments, eliminating the restriction that we start and end at the assassins' headquarters. Screenshot from *Assassin's Creed Series.* © Ubisoft Entertainment. All Rights Reserved. *Assassin's Creed,* Ubisoft, and the Ubisoft logo are trademarks of Ubisoft Entertainment in the U.S. and/or other countries.

any closer to finding an algorithm to solve TSP). For one, it doesn't make much sense that Ezio would need to return to his headquarters after collecting the eleventh fragment. And since the player doesn't necessarily always *start* at the assassin's headquarters, this assumption isn't strictly necessary either. In particular, the restriction that the path be *closed*—that is, that you return to your starting point—doesn't seem like a particularly natural one.

If we eliminate that restriction, we get a slightly different path, though not a different problem in general (Figure 6.13). This is sometimes referred to as an *open tour* for the traveling salesman, though it's just as hard as the original problem.[17]

In this case, however, an open tour does result in a shorter minimum path—in fact, this path is only 76% as long as the first one!

But wait! There's another improvement we can make. As in most open-world games, *Assassin's Creed: Revelations* uses a "fast-travel" mechanic meant to help cut down on time spent moving from one location to the next. In this case, Ezio can use underground tunnels to instantly travel from one part of the city to another.

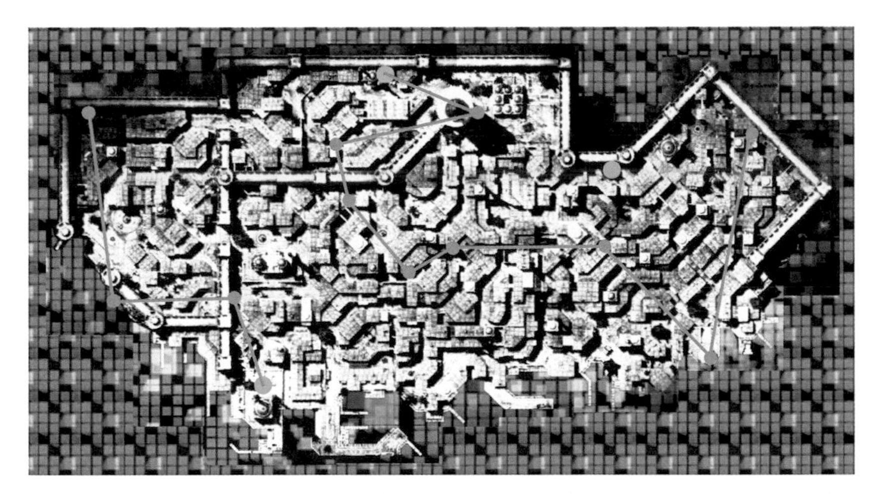

FIGURE 6.14. Shortest path to the eleven data fragments, eliminating the restriction that we start and end at the assassin's headquarters. The green dots indicate fast-travel locations. Screenshot from *Assassin's Creed Series*. © Ubisoft Entertainment. All Rights Reserved. *Assassin's Creed*, Ubisoft, and the Ubisoft logo are trademarks of Ubisoft Entertainment in the U.S. and/or other countries.

Galata has three tunnel locations, including one in the assassin's headquarters. In some cases, this means that the shortest distance between two data fragments isn't the straight line between them, but instead is a line through one of these fast-travel locations. By accounting for these, we can trim a bit more off the minimal distance.

The path in Figure 6.14, which includes one fast-travel leg via underground tunnels, has around 98.7% the length of the path in Figure 6.13.[18] Of course, this ignores the fact that in the actual game, there's a load time involved when fast-traveling from one location to the next; the change in scenery doesn't happen instantaneously. If you're trying to minimize time instead of distance, the difference between this path and the other open one may be negligible, or possibly even worse.

Note that when considering fast travel, calculating distances can be a little tricky, since the distance between two cities isn't necessarily the straight line between them. So, we first need to ask our computer to find all the distances between each pair of data fragments, and then use those data to find the shortest path.

For just eleven fragments, this is no big deal. But fairly early on, the game world expands well beyond Galata. In the main area of

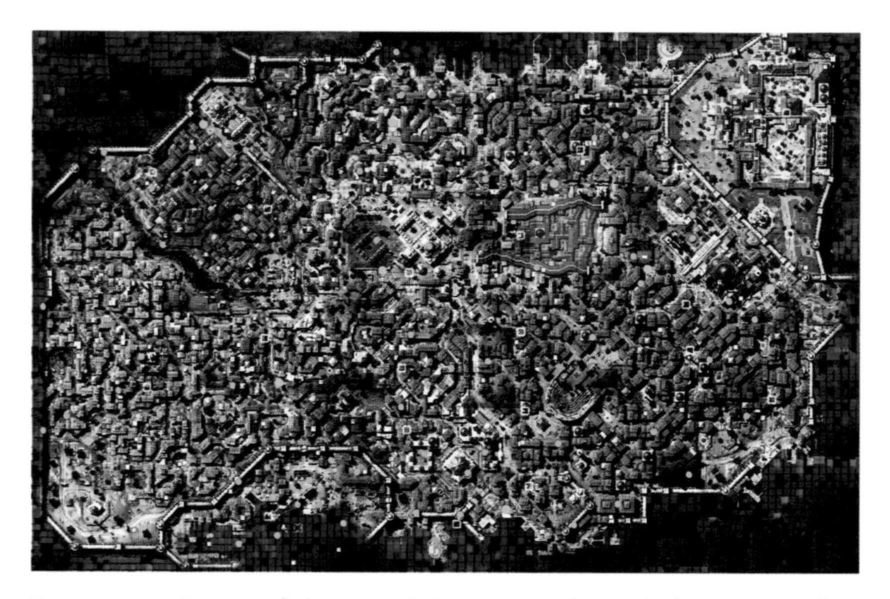

FIGURE 6.15. A map of the rest of Constantinople. As before, orange dots indicate data fragments and green dots indicate tunnels for fast travel. Can you find the shortest path to all the fragments? Screenshot from *Assassin's Creed Series*.

Constantinople, there are seven times as many data fragments—seventy-seven in all[19]—and another ten underground tunnels that can be used for fast travel. After calculating all the distances, there are still seventy-seven options for the first data fragment to collect, seventy-six options for the second, and so on (Figure 6.15). This means that the total number of paths within the main area of Constantinople is equal to $77!$, or around 1.43×10^{113}.

It's also worth pointing out that the fast-travel rules can vary from game to game, even within the same series. Starting with *Assassin's Creed III*, players could travel directly to certain points on the map at any time, regardless of their current location. In other words, the distance from your current location to some points on the map is always zero.

In particular, *Assassin's Creed III* takes place in America during the eighteenth century, and a large portion of the game world is the frontier.

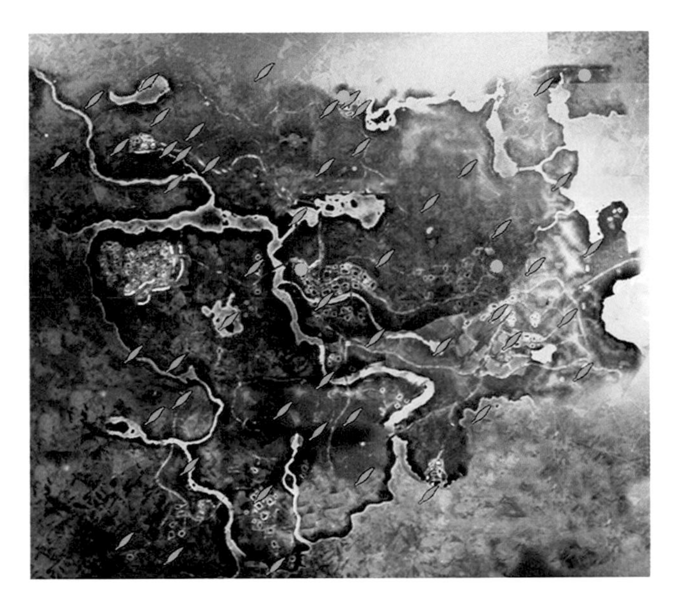

FIGURE 6.16. What do you think the shortest path to all of these feathers might look like? Screenshot from *Assassin's Creed Series.* © Ubisoft Entertainment. All Rights Reserved. *Assassin's Creed,* Ubisoft, and the Ubisoft logo are trademarks of Ubisoft Entertainment in the U.S. and/or other countries.

In this area, there are fifty feathers that the player can collect, as well as four locations on the map that the player can instantly travel to (Figure 6.16). How do you think the existence of these locations might influence the shape of the shortest path to collect all of the feathers?

And in the race to make games bigger, more complex, and filled with more *things*, the *Assassin's Creed* series has continually proved up to the challenge. Later games in the series featured naval exploration, which expanded the game world considerably. And even when the open world became less open—as in the transition from *Assassin's Creed IV*, which let you explore the entire Caribbean, to *Assassin's Creed: Unity*, which restricted you to just the city of Paris—the game designers compensated by increasing the number of things to do per square kilometer of game world (Figure 6.17).

In fact, 2015's *Assassin's Creed: Syndicate* had so many things to do, you couldn't see them all at once on the map. Only by zooming in to a specific location could you identify all the points of interest. With the entire map in view, only a few key features were present. And

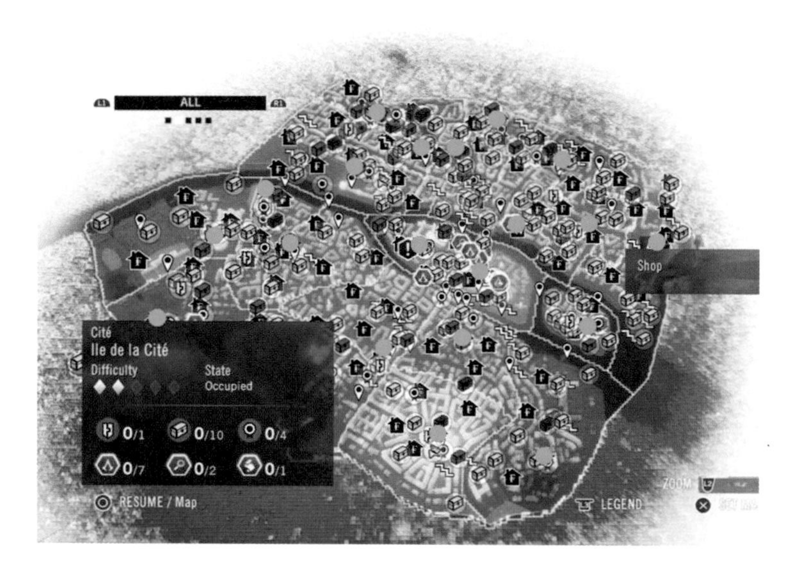

FIGURE 6.17. In *Assassain's Creed*: Unity, the player can travel to any of the green dots at any time, and from any current location. To some, this map may look like fun. To me, it's a little overwhelming. Screenshot from *Assassin's Creed Series*. © Ubisoft Entertainment. All Rights Reserved. *Assassin's Creed*, Ubisoft, and the Ubisoft logo are trademarks of Ubisoft Entertainment in the U.S. and/or other countries.

even if you could see everything at once, finding the most efficient route in the game was further complicated by the fact that one of the fast-travel points was on a moving train. As you can imagine, trying to solve TSP when the locations themselves are also moving can get rather complicated.[20]

Even with stationary cities, as the number of destinations grows, finding the shortest path between them becomes harder. While there are improvements to the naïve "search through all paths to find the shortest one" approach indicated here, the number of calculations grows very rapidly with the number of points one wants to connect. Nowadays it's possible to solve TSP on networks of more than 10,000 points, though the amount of computation time required is substantial.[21, 22]

When it comes to **P** vs. **NP**, maybe completing fetch quests isn't a total waste of time. Though they are frequently tedious and unnecessary, they may also provide a testing ground for any potential **NP** problem-solving algorithms.

6.7 CLOSING REMARKS

Through this tour of **NP** problems in games, one question persists: Why do **NP** problems keep showing up? In some sense, this may be more of a philosophical question than a mathematical one. Perhaps it has something to do with the types of games that we find enjoyable. When describing *Tetris* in his book *Fun Inc.: Why Gaming Will Dominate the Twenty-First Century*, author Tom Chatfield writes,

> The significance of presenting such a complex problem so accessibly is in the degree to which it raises the boredom threshold of the player. ... Playing *Tetris* is a mathematically endless undertaking. You can never say you have mastered it in terms of exhausting its possibilities: you can only improve your tactics. Moreover, it is a mathematical inevitability that even the greatest player is eventually doomed to lose (Chatfield 2011, p. 41).

In other words, we appreciate—possibly even require—a certain amount of complexity in our games. Make a game too simple, and it can't hold our attention (just ask anyone who's played the card game War). Make it too complex, and it won't be comprehensible, much less fun to play. Perhaps **NP** problems continue to appear because they represent a sweet spot in complexity that makes for compelling gameplay.

Will someone solve one of the most famous open problems in mathematics by sitting on her couch and playing video games all day? I think it's safe to say that the answer is no. But these video games may serve as a worthy introduction. While playing these games probably won't make you a famous mathematician, studying them more thoroughly just might.

7
The Friendship Realm

7.1 TAKING IT TO THE NEXT LEVEL

One of my all-time favorite shows is *Flight of the Conchords*. Although it lasted only 22 episodes, what the show lacked in longevity it made up for in awkwardness and heartwarming folk-comedy music. The semi-autobiographical show centered on Bret and Jemaine, New Zealanders searching for fame and fortune in New York City. It also launched the real-life versions of Bret and Jemaine to a new level of success: since its run, Jemaine has appeared in a number of films, and Bret won a Best Original Song Oscar for the 2011 film *The Muppets*.

For the characters in the show, however, success always seemed elusive. Bret and Jemaine lived in a small apartment and struggled to find work. The task of finding them gigs fell upon their manager, Murray, whose enthusiasm for the job far surpassed his ability.

Murray's intentions were pure, however, and his affection for Bret and Jemaine was impossible to deny. Consider, for example, the following exchange from the second season of the show, in an episode titled "Murray Takes It to the Next Level" (Bobin et al. 2009):

MURRAY: Now, we've known each other for quite some time in the professional realm. I'd like to push things forward in the friendship realm.

JEMAINE: What's the friendship realm?

MURRAY: Well, you've heard of a realm.

BRET: Mmm.

MURRAY: Yeah?

JEMAINE: Yes.

MURRAY: Well, this is like a friendship one. A group of people, getting together, basically calling each other friends. Look at this.

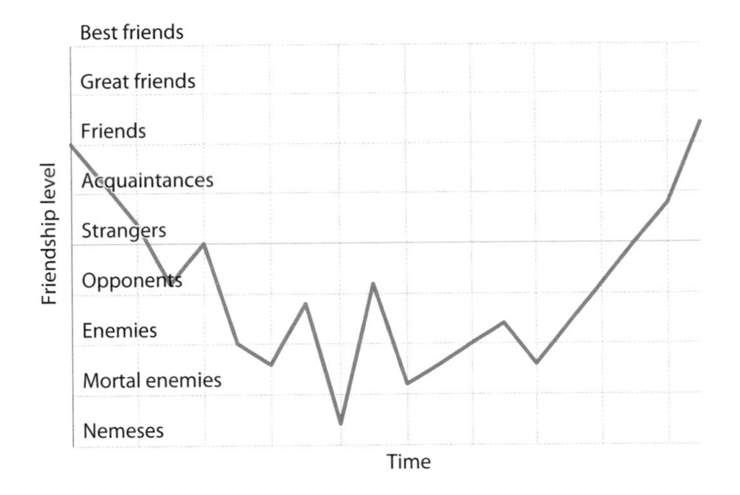

FIGURE 7.1. The people in this relationship start as friends, but things devolve until they are almost nemeses before the situation improves.

Murray then shows Bret and Jemaine his model of friendship, which he represents as a graph along two axes: friendship and time. The graph also has a number of friendship thresholds. Once a relationships crosses one of Murray's thresholds, it transforms from one category to another. He describes the graph as follows: "If you have a look along here on the x-axis, this represents time passing, and on the y-axis here, this is the different levels of friends."

Admittedly, Murray's model is fairly qualitative and highly subjective (how he assigns a numerical value to the actions of others is anyone's guess). The general idea of trying to model friendship, however, is an admirable one and is ripe for some mathematical tinkering. If you wanted to, you could apply this idea to your own friendships. For example, if you and a friend go through a somewhat rocky period in your relationship, your graph over time might look something like Figure 7.1.

But what does this have to do with video games? Well, to a certain extent, Murray's model is not so different from ones that have already appeared in some games. And even though games are getting better and better at modeling physical reality, their ability to model the thornier nature of *human interaction* is still, in many ways, not so different from what Murray does in his personal life.

In this chapter, we'll take a look at how video games have explored friendship and try to understand some of the underlying mathematics. In the process, we'll examine several different mathematical models for relationships. Some of the models are more complex than others, but we'll see that additional complexity isn't always a good thing. Instead, there's often a tradeoff between the complexity of the model and how widely applicable it is. This, in part, is what makes modeling so rich and so challenging. Though more accurate models of relationships may not translate into better games, at the very least they could help Murray (and the rest of us) refine our thinking when it comes to how our social bonds change over time.

7.2 FRIENDSHIP AS GAMEPLAY: *THE SIMS* AND BEYOND

It's not difficult to conjure up images of avid game players as isolated loners, mindlessly mashing buttons and devouring Hot Pockets. But at its best, gaming is a social phenomenon—why else would consoles come with multiple controller ports? And while friendship has long been a part of the experience of playing games, it hasn't always been part of the gameplay itself.

Historically, role-playing games (RPGs) tended to explore friendship the most deeply at first, due to their overall emphasis on story and character. Games like *Chrono Trigger* and *Final Fantasy VII* featured large casts of playable characters whose relationships helped propel the story.

The relationships in those games are somewhat influenced by choices the player makes. For example, both *Chrono Trigger* and *Final Fantasy VII* let you form a party with three different characters; this active party is the one you use to explore the world and take into battle. However, in *Chrono Trigger* there are seven playable characters; in *Final Fantasy VII*, there are nine. In other words, at any given point in either game, a majority of the playable characters are on the sidelines. Players must continually make choices about whom they want in their parties.

Friendship in these games served more than just a narrative purpose, however. In many role-playing games at that time, social bonds had a gameplay function as well. Different characters had different attributes, and so different parties catered to different play styles.

More importantly, active members of the party became stronger at a faster rate than nonmembers, as measured by the RPG staple known as experience points (which are gained by winning battles). In *Chrono Trigger*, nonactive members of a party gained only 75% of the experience points that active party members did; in *Final Fantasy VII*, nonactive members gained only 50%. In other words, these games created a strong gameplay incentive to stick with a team if your goal was to become strong as quickly as possible. Moreover, the longer you stuck with one team, the less incentive there was to switch the team up, since the nonactive players would have been comparatively much weaker.

Ties between characters may have emerged as a result of these types of gameplay rules in classic role-playing games, but in none of them is friendship itself a central game mechanic. The most well known series to turn the process of making friends into a game is *The Sims*. Now the best-selling franchise for personal computers in history, the series has spawned several sequels and has millions of fans from all over the world. The series has certainly evolved since its first iteration, but even the most recent entries share quite a bit with their predecessors.

Each game in the series allows you to create an avatar—the titular "Sim"—and control that Sim in a virtual community. As a (hopefully benevolent) overlord, you have a fair amount of responsibility: not only are you in charge of your Sim's larger life goals, such as his career path and how he behaves in his love life, but you also need to pay attention to the smaller things, like personal hygiene and paying bills.

Among your many responsibilities, you must also help your Sims enrich their social lives by growing and maintaining their friendships. Compared to real life, this is fairly straightforward: your Sim can easily approach strangers, make small talk on the phone, and always has a good joke up her sleeve. Though it's possible for your Sim to elicit an unfriendly reaction (few Sims like braggarts, and flirting with strangers is frowned upon), it's generally not too difficult to make friends in the game.

What does it mean, though, to make a friend? We all know who our friends are (or, at least, we hope we do), but if we look back on a relationship in hindsight, it can be difficult to pinpoint the moment when friendship blossomed between two acquaintances. Not so in

FIGURE 7.2. Virtual romance, portrayed in *The Sims 4*. © EA games

the world of *The Sims*, though; in this game, much like in Murray's mind, friendship is a binary state that is activated when the positive interactions between two people cross a particular threshold.

Originally, these measurements were overtly quantitative. If two Sims had an enjoyable conversation, their relationship meters would increase by some numerical value that you could see. Conversely, if the actions of another Sim offended your own, your Sim's relationship meter would decrease by some value. A value of zero corresponded to a neutral relationship, neither friend nor foe. If each person's meter for the other went above fifty, the people became friends, and if it dipped below minus fifty, they became enemies.

By the time *The Sims 2* was released in 2004, the relationship model had become a bit more sophisticated. This time around, every one of your Sims' relationships was modeled with *two* meters: one to measure short-term feelings and one to measure the long-term durability of the relationship. This latter meter responded much more slowly to changes in the relationship, both positive and negative.

Friendship meters in *The Sims 3*, released in 2009, were much simpler. In this iteration, there was only one meter, and no numbers at all. Instead, the proportion of the meter that was shaded green (or red, in the case of dislike) was the only indicator of friendship. There were also icons to indicate the type of friendship: family, coworkers, and so on.

The Sims 4, released in 2014, reinforced this more qualitative approach. The most significant change here was the separation of

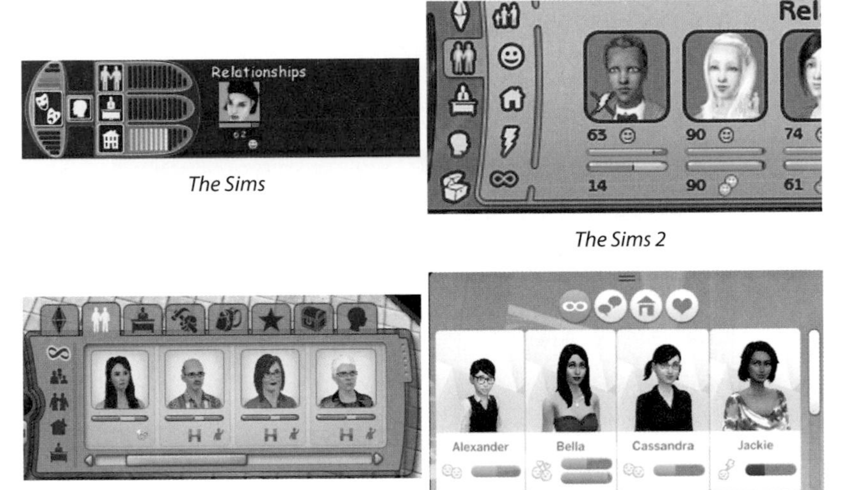

FIGURE 7.3. The evolution of friendship in *The Sims*. In the original game, the player's Sim thinks of the female character as a friend, since the relationship score is greater than fifty. In *The Sims 2*, each relationship had two components: short-term relationship strength on top, long-term strength on bottom. Friendship information in *The Sims 3* was less quantitative and used icons to indicate different types of friends. Finally, *The Sims 4* offered up a more streamlined design and added a romance bar independent of the friendship bar. © EA games

friendship and romance into their own separate meters. In previous versions of *The Sims*, it was easy to parlay a friendship into a romantic relationship, because there was little separation between the two. With *The Sims 4*, however, it became possible to more accurately model purely platonic friendship.

With some small variations, the general idea behind measuring friendship in these games is really not so different from Murray's model. For instance, both models use a tiered system of friendship. Positive interactions yield increases in friendship meters, and negative interactions lead to decreases. Cross a certain threshold, and suddenly you advance to the next level of friendship status. And though *The Sims* no longer explicitly provides numbers, it would be easy enough to assign numerical values to the relationship based on what percent of the friendship meter is full.

FIGURE 7.4. Michele Dachss, one of many buddies you may encounter while playing *Far Cry 2*. Screenshot from *Far Cry 2*. © Ubisoft Entertainment. All Rights Reserved. Far Cry, Ubisoft and the Ubisoft logo are trademarks of Ubisoft Entertainment in the US and/or other countries. Based on Crytek's original Far Cry directed by Ceval Yerli. Powered by Crytek's technology "CryEngine".

The Sims isn't the only series to model friendship in this way. In *Far Cry 2*, for example, players take on the role of a mercenary sent into a war-torn African nation. Through the course of the campaign, the player meets other mercenaries, known as "buddies." Helping a buddy during missions increases your buddy score and also unlocks certain gameplay upgrades. Conversely, if you fail to help him or her, your buddy score may decrease. At any point in the game, the mercenary with the highest buddy score is listed as your best buddy, and the mercenary with the second highest score is your second best buddy. Both of them will offer you assistance as you explore the game world.[1]

Recent entries in the open-world series of games called *The Elder Scrolls* incorporate similar models. In *The Elder Scrolls V: Skyrim*, characters with whom the player interacts are each assigned one of the "disposition" scores in Table 7.1.[2] Characters' disposition scores change in response to the players' actions in the game world. As you might expect, a character treats the player very differently if his or her disposition score is high than if it is low.

7.3 A GAME-INSPIRED FRIENDSHIP MODEL

Though these friendship models may have striking similarities despite spanning a variety of games, some are more realistic than others. One

TABLE 7.1. Friendship, as Measured in *Skyrim*

Disposition Score	*Name*
−4	Archnemesis
−3	Enemy
−2	Foe
−1	Rival
0	Acquaintance
1	Friend
2	Confidant
3	Ally
4	Lover

interesting feature of friendship in every version of *The Sims* that is absent from many other games is friendship decay. No matter how strong the relationship, if your Sim doesn't hang out with her friends, or at the very least talk to them on the phone, the friendships will eventually deteriorate. In other words, if left unchecked, all friendship scores tend toward zero.

This phenomenon exists in real life and has even been given a name: the second law of thermodynamics for sentimental relationships.[3] This law essentially states that people need to put effort into relationships in order to sustain them, and without that effort the quality of the relationship is bound to erode.

Let's take a closer look at this idea within the context of *The Sims*, by exploring a variation of the game's friendship model. To start, we'll denote the friendship score between two people t days after they meet by the function $f(t)$. Larger function values correspond to stronger friendships, and smaller function values correspond to weaker friendships. For simplicity, we'll assume that this score describes both individuals' feelings toward the other. In other words, we ignore the possibility that one person's strong feelings of camaraderie aren't reciprocated; for the time being, all feelings are mutual.

In keeping with what we observe in *The Sims*, we'll assume that these friendships naturally decrease in strength by a constant factor, say r, each day. For instance, a decay factor of $r = 0.5$ for a given friendship would mean that, left unchecked, the friendship would deteriorate by 50% each day. (This doesn't sound like a particularly robust relationship.) In order to counteract this friendship decay, the relationship needs to be strengthened by some form of interaction. Here we'll make one more simplifying assumption: let's assume that this strengthening is proportional to the amount of effort that is put into the friendship. When more effort is put in, the friendship grows; and if no effort is put in, the friendship wilts. We'll let $c(t)$ denote the amount of effort put into the relationship on day t.

Combining all of our assumptions, we are led to the following equation defining f:

$$f(t+1) = r \cdot f(t) + a \cdot c(t), \tag{7.1}$$

for some fixed numbers r and a. Since we've assumed that effort put into a friendship causes it to strengthen, a must be nonnegative. Also, since r represents the decay factor, we must have $0 < r < 1$.

Notice also that when $c(t) = 0$ (that is, when no effort is made), the relationship will naturally decay each day by a factor of r. When r is small, the relationship will deteriorate quickly, while if r is close to 1, it decays more slowly. Similarly, the effect of the effort spent on the friendship is amplified by a, so when a is large two people don't need to put much effort into the friendship in order to maintain (or improve) it. In summary, it's easier to keep a friendship when r and a are both large.

Equation (7.1) is an example of a *recursive* equation, since the friendship score on any day is expressed *recursively* in terms of the friendship score on the previous day. However, with a little exploration we can derive an *explicit* formula for $f(t)$, one that doesn't require us to know $f(t-1)$. Indeed, if we look at some small values of t, we see that

$$\begin{aligned}
f(2) &= r \times f(1) + a \times c(1), \\
f(3) &= r \times f(2) + a \times c(2) \\
&= r \times (r \times f(1) + a \times c(1)) + a \times c(2) \\
&= r^2 f(1) + a(r \times c(1) + c(2)), \\
f(4) &= r \times f(3) + a \times c(3) \\
&= r \times (r^2 f(1) + a(r \times c(1) + c(2))) + a \times c(3) \\
&= r^3 f(1) + a(r^2 \times c(1) + r \times c(2) + c(3)),
\end{aligned}$$

and in general

$$\begin{aligned} f(t+1) &= r^t f(1) + a(r^{t-1}c(1) + r^{t-2}c(2) + \cdots + c(t)) \\ &= r^t f(1) + a\sum_{i=1}^{t} r^{t-i} \times c(i). \end{aligned}$$

This formula isn't particularly elegant, but it does allow us to compute the friendship score on any day if we know the initial friendship score ($f(1)$), the amount of effort the pair has spent on the friendship each day, and the values of the parameters a and r.

Even so, that's a lot of information to keep track of, especially when t (the number of days) is large. Thankfully, in certain situations a precise formula may be more than we need, and we can replace the above equality with some approximation.

7.4 APPROXIMATIONS TO THE MODEL

Suppose that two people have an initial level of familiarity $f(1)$. What can our formula tell us about how their friendship might look in a year, or ten years? And how can we simplify the formula without sacrificing too much of the model?

We've seen that with our current model, on day $t + 1$ the friendship score will be

$$f(t+1) = r^t f(1) + a\sum_{i=1}^{t} r^{t-i} \times c(i).$$

Remember that $0 < r < 1$, since r represents a friendship decay factor. Because of this, as t gets larger, r^t approaches 0. Therefore, when t is large, $r^t f(1)$ is approximately zero, and so we have the estimate

$$f(t+1) \approx a\sum_{i=1}^{t} r^{t-i} \times c(i). \tag{7.2}$$

This already gives us our first interesting finding: the long-term behavior of the friendship is independent of the initial friendship score! This should make some intuitive sense: after all, the set of friends you have now isn't necessarily a great predictor for your set of friends in a decade or two. We continue to make friends throughout our life, and even in our highly connected culture we often lose touch with older friends. The current friendship score between two people, no matter how high or how low, isn't necessarily a great predictor for how much longevity the friendship has.

In order to push our model further, we need to wrestle a bit with the sum on the right-hand side of expression (7.2). In general, the sum isn't so nice, since we don't know anything about the values $c(i)$; however, by analyzing certain special cases we can reveal some general phenomena.

The simplest possible case would be if the two people whose friendship we're modeling put the same amount of effort into the friendship every day. In this case, we'd have $c(1) = c(2) = \ldots = c(t)$, and so we could set these common values equal to a single constant, say c. We could then rewrite expression (7.2) as

$$a\sum_{i=1}^{t} r^{t-i} \times c(i) = a \times c \sum_{i=1}^{t} r^{t-i}.$$

Now we have a sum just over consecutive powers of r. This type of sum is given a special name (a *geometric* series), and in fact we can simplify the sum considerably. In this case, we have[4]

$$\sum_{i=1}^{t} r^{t-i} = \frac{1-r^t}{1-r},$$

which is approximately equal to $\frac{1}{1-r}$, since again $r^t \approx 0$ for t sufficiently large.

Putting everything together in this scenario gives us the approximation

$$f(t+1) \approx \frac{ac}{1-r}$$

when t is sufficiently large. In fact, in the limit we would have an exact equality:

$$\lim_{t\to\infty} f(t) = \frac{ac}{1-r},$$

although technically this doesn't make sense, since nobody's time on this Earth is infinite.[5]

Nevertheless, the expression $\frac{ac}{1-r}$ gives us an estimate for the long-term strength of a friendship, assuming that the two parties spend a roughly constant amount of effort on the relationship. We'll call the value of this expression the *long-term friendship score*.

The long-term friendship score has many properties that agree with our intuition. For example, when $c = 0$, we see the long term-friendship score is also 0, agreeing with the idea that two friends who never spend time together will gradually drift apart. Moreover, the long-term friendship score increases as a and c increase, which also makes sense. Certainly the more effort two people put into a friendship, the stronger their relationship should be. At the same time, large values of a compensate for smaller values of c, so even if two people don't get to spend a lot of time together, they can still have a high long-term friendship score if the quality of their time spent together is high (in other words, if their a value is large).

Similarly, as r approaches 1, the denominator in the long-term friendship score approaches 0, so that the long-term friendship score itself gets larger. This too agrees with our intuition, since an r value very close to 1 means that the friendship exhibits very slow decay over time.

Finally, we can bring back our initial friendship score $f(1)$ and ask whether or not in the long run a given friendship will grow or decay. In order for growth, the following inequality must be true:

$$f(1) < \frac{ac}{1-r}.$$

In order for decay, the opposite inequality must hold.

Of course, it's not entirely clear how one would go about measuring any of these quantities. While this analysis may be an interesting exercise, it remains a relatively abstract description of how friendship may change over time.[6]

7.5 THE COST OF MAINTAINING A FRIENDSHIP

For all the mathematics we've been able to squeeze out of our model, the discussion so far has been missing a crucial component. Murray's life choices aside, we typically don't put effort into our friendships because we want to increase some abstract "friendship score." We put effort into our friendships because we like to and because it makes us happier! In other words, we do it because effort put into relationships increases our *utility* (or at least, we expect it to).

How might our utility be affected by the time we spend with friends? There are a few things to consider. First, it seems reasonable to assume that we gain more utility from spending time with close friends than acquaintances, even though the exact relationship between utility and friendship score may be hard to pin down. On the other hand, there's a limit to how much utility we gain from putting effort into the friendship. Even best friends may start to get a little sick of each other if *all* of their energy goes into the friendship. Also, if you're spending too much time with one person at the expense of life's other obligations, this may have a negative effect on your overall quality of life.

The Sims understands this phenomenon implicitly. In addition to measuring friendship strength, it also has a model of each Sim's utility. Like friendship, utility is measured graphically with a meter (called the aspiration bar) that fills up when the Sim is feeling fulfilled. When the meter is green, the Sim's utility is high, and when it's red, utility is low. Moreover, utility is decomposed into specific categories, so that the player can see what actions are most likely to improve her Sim's overall level of satisfaction.

For example, consider the meters in Figure 7.5, which shows a sort of utility dashboard for a Sim at two moments in time. In both images, the meter on the left represents overall utility: the greener it is, the happier the Sim. On the right are six other meters, representing some component of the Sim's utility.

In the top image, each of the six meters on the right is green, meaning that in every category the Sim is relatively satisfied. As a consequence, the Sim's overall utility meter is, unsurprisingly, fairly green.

Imagine, though, that the Sim then spends an entire day talking with a friend, at the expense of everything else. At the end of the day, the Sim's "social" meter will be quite high, but if the Sim neglects to eat, shower, sleep, or use the bathroom, his overall utility will suffer. The second image depicts this situation. Of course, in real life, few of us would neglect to do all of these things when spending time with

FIGURE 7.5. Sim utility meters as seen in *The Sims 3*. © EA Games

someone. It's still possible for the time we devote to a friendship to negatively affect our utility, however. For example, suppose you offer to help your friend clean up after a party at her place. Even though this will likely bolster your friendship, it may put a damper on your overall happiness. You may not get enough sleep that night, or you may not be able to take care of some other errand that you had planned on doing.

The point is simply that we can model our utility from a friendship in two dimensions. One of them, call it $U(f)$, is the utility you gain from the friendship score itself—the closer you are to a friend, the more happiness that friend will bring to your life. The other component, call it $D(c)$, represents the *disutility* that arises if too much effort is spent on the friendship (Figure 7.6). Though D may be negative for small values of c (i.e., it may represent utility rather than disutility), eventually D will become increasingly positive. In this way, the overall utility you gain from the friendship can be written as the difference $U–D$.[7]

If our function D starts off decreasing with c but eventually increases, then (under some reasonable assumptions on D)[8] we know that D has a minimum value. In other words, there is a daily amount of effort c^* that you should invest in your friendship in order to minimize your disutility. If the time you spend with your friend is greater than c^*, D will begin to increase.

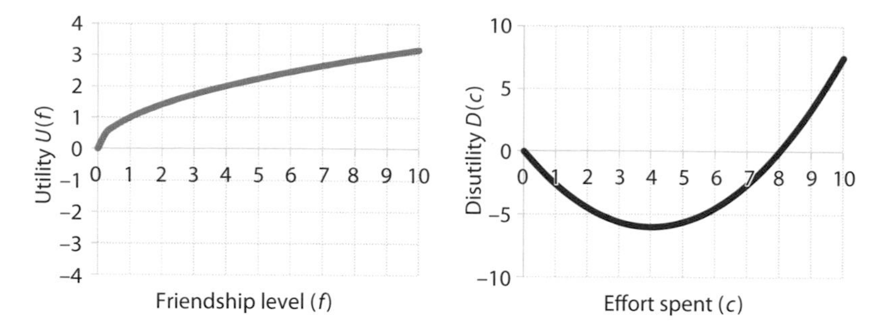

FIGURE 7.6. On the left, possible behavior for the function $U(f)$. Notice that U increases with f, but the rate at which it increases levels off. On the right, possible behavior for the function $D(c)$. Since D initially decreases and then increases, it must have a minimum value. (Images adapted from Rey 2010).

So, if we want a friendship to bring us happiness, we should figure out the amount of effort that minimizes our disutility, and then invest that amount of effort in the friendship. Right?

Well, not quite. It turns out that simply *minimizing* your disutility from effort spent may not *maximize* the overall utility from the friendship. In fact, it may be that in order to be as happy as possible, you need to put more effort into the friendship than you'd otherwise choose to. In other words, friendship isn't free: it requires effort to maintain, and that effort may be beyond the amount you'd ideally choose to spend. The benefit, of course, is that you're rewarded for the effort with a stronger friendship.

Still, you may not be willing to put more effort into a friendship once the amount of effort starts increasing your disutility. But what if the friendship is a central component to your life? What if, for example, the friendship is between you and your spouse?

7.6 FROM VIRTUAL FRIENDS TO REALISTIC ROMANCE

The model of friendship and utility we've considered here was first described in a 2010 paper titled "A Mathematical Model of Sentimental Dynamics Accounting for Marital Dissolution," by professor JoséManuel Rey of the Universidad Complutense in Madrid. Rey's paper is concerned with romantic relationships, not just friendships, and in keeping with his paper, let's now shift our focus outside of the "friend zone." Although the relationship dynamic is different, Rey's model is based on the same assumptions we've already discussed:

1. The relationship is symmetric, in that person A's relationship score for person B is the same as person B's relationship score for person A, and both people derive the same utility from spending time in the relationship. In other words, the fundamental unit is the pair, not the individual. (This is called the *weak homogamy* assumption.)
2. The relationship will naturally decay if effort is not put into maintaining it. (This is called the second law of thermodynamics for sentimental relationships.)

Rey's model is more sophisticated than the one we discussed above, because he uses a continuous time model. That is, instead of looking at a friendship score (or relationship satisfaction score) on a day-by-day basis, his model considers the score for any time t. In particular, this means that instead of an equation like equation (7.1), his assumption (2) yields a *differential* equation of the form

$$f'(t) = -r \times f(t) + a \times c(t).$$

If your calculus is a little rusty (or a little nonexistent), the left-hand side is called the derivative of $f(t)$ and measures the rate of change of f. In other words, it tells us how quickly or slowly the relationship is changing at time t. This equation tells us that this rate of change has two components: it is increased by time spent together (represented by the $a \times c(t)$ term), but it also naturally decays in proportion to the relationship score. This is why there is now a minus sign in front of the $r \times f(t)$ term; we're now measuring a rate of change in the friendship score, not the friendship score itself. If $r \times f(t)$ didn't have a minus sign in front of it, the rate of change would always be positive, so that the friendship score would always be increasing.

Because of an improved model,[9] Rey is able to reach a more precise version of our conclusion from before. In particular, Rey proves the existence of a relationship state in which both parties are happy and the relationship is in equilibrium. However, this steady state depends on the willingness of each party to expend more effort into the relationship than they'd ideally choose to. Rey deems this difference the *effort*

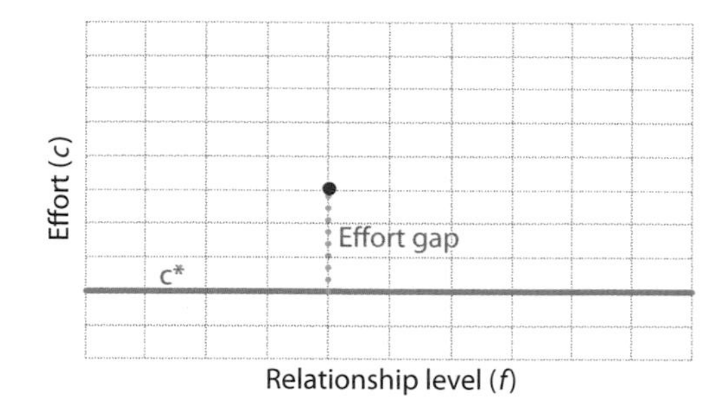

FIGURE 7.7. Rey proved the existence of a point on the $f - c$ axis that maximizes a person's utility (indicated by the red dot above). However, the effort required to achieve that utility will exceed the value c*—in other words, to maintain marital bliss, you'll need to put effort into the relationship beyond what minimizes your disutility from effort spent. (Image adapted from Rey (2010).)

gap; provided the gap is not too large, the couple can, in theory, remain happy forever. Fair warning: the equilibrium relationship state is unstable, meaning that it can derail itself at any time if either individual doesn't put in the right amount of effort.

As Rey points out in his paper, this model helps to explain an apparent paradox of modern romance: while most people enter into a serious relationship believing that it will last forever, the divorce rate in many countries is still quite high. Viewed from the model's perspective, however, this is not so surprising. Indeed, the model explains why, despite a couple's best intentions, the relationship may not succeed. Rey shows that if the effort gap is too large for the individuals to bridge, then the relationship will eventually deteriorate.

In other words, relationships take work. More work than you might otherwise choose to put into them. But if the reward is a long-lasting and happy relationship, isn't that worth the investment?

7.7 MODELING DIFFERENT PERSONALITIES

Your answer to the question above undoubtedly depends on how you feel about long-term relationships in general. If you have a fear of commitment, for example, maybe the thought of spending a lifetime with one person may terrify you. But if you're a hopeless romantic, the amount of effort you are willing to put into a relationship may be limitless.

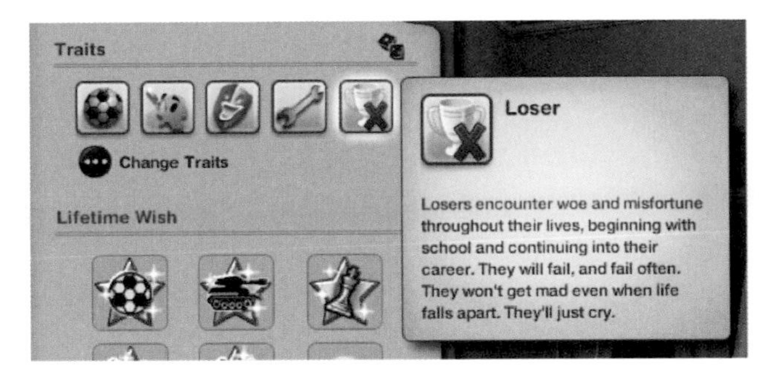

FIGURE 7.8. One possible combination of five traits in *The Sims 3* is pictured above. Sadly, the Sim represented here will most likely be unlucky in life. © EA Games

These disparities aren't captured in our model because we assumed that both people in a friendship or relationship had essentially the same personality (see the weak homogamy assumption above). While it's certainly the case that people in relationships frequently have similar personalities, it's also worth exploring the adage of "opposites attract."

The assumption of weak homogamy seems like an unrealistic one to make in general, and it's certainly not a valid way to model gameplay in *The Sims*. For one thing, it's simply not that true that two Sims always have the same friendship score: my Sim could view your Sim as a good friend, for example, while your Sim might consider mine only an acquaintance. In fact, your Sim might not like mine at all!

Moreover, just like real people, Sims have a wide array of personality types. In fact, starting with *The Sims 3*, the player can specify certain personality traits for any character they create. Different combinations of traits yield different personalities (Figure 7.8).

Differences in personality yield different relationship dynamics. For example, suppose a Sim with the "hopeless romantic" trait falls for one with the "commitment issues" trait. This isn't necessarily a disaster waiting to happen, but the fundamental difference in personality does mean that managing the relationship will be more difficult for the player than it might otherwise be.

Hopeless romantics respond more positively to the advances of others (both in *The Sims* and, arguably, in real life). While the hopeless romantics might dive into marriage with someone they don't know that well, Sims with commitment issues need an extremely high friendship score in order to accept or offer a proposal.

Many of us have also seen or experienced real-life examples of relationships that suffer from similar personality disparities. Hopeless romantics can fall hard for people who may rebuff their advances out of a fear of getting serious too quickly. Indeed, if one person comes on too strong, it's entirely possible for his growing feelings of attraction to cause his partner's feelings to diminish.

These dynamics are ripe for mathematical modeling. Perhaps no one knows this better than Steven Strogatz, a mathematics professor at Cornell University. In a short note titled "Love Affairs and Differential Equations," Strogatz proposed relationship dynamics as a source of inspiration for studying some standard problems in differential equations (1988). We already saw one example of a differential equation in this chapter; let's take a look at how we can use more equations of this type to model a relationship between two people with different personalities.

We'll start with a relatively simple scenario. Imagine a man (we'll call him Mario) has feelings for a woman (we'll call her Peach). She's a romantic, but she's also a realist, and while her feelings for Mario can get fairly intense, she only wants to be with him when he wants to be with her. Mario, on the other hand, has a bit of a fear of commitment, and when things get too intense he tends to back off. In other words, his feelings for Peach decrease when her feelings increase. But if Peach's feelings decrease, Mario's feelings kick back into gear, and suddenly he wants to be with Peach.

We can model Mario's feelings for Peach and Peach's feelings for Mario using two functions, $M(t)$ and $P(t)$, respectively. Based on our description, it's reasonable for our equations to satisfy the following differential equations:

$$M'(t) = -a \times P(t),$$
$$P'(t) = b \times M(t),$$

for some positive constants a and b. Remember that on the left we have derivatives, which measure rates of change. The minus sign in the first equation tells us that Mario's feelings change in opposition to Peach's feelings—when she's hot, he's cold, and vice versa. On the other hand, Peach's feelings change in parallel with Mario's.

Qualitatively, you can probably imagine what happens. Let's say Mario and Peach start off with strong feelings for each other. Because of Peach's strong feelings, Mario will withdraw, and his feelings will decrease. Since Peach's feelings mirror Mario's, she too will withdraw, but as she becomes less interested, Mario will become more interested. At some point, Mario will begin to turn on the charm, at which point Peach's feelings will begin to grow, and the cycle will repeat.

With this understanding in mind, it's not so surprising that we can describe $M(t)$ and $P(t)$ more explicitly. They're nothing more than sums of trigonometric functions! Indeed, it's possible to show that $M(t)$ and $P(t)$ must have the general form[10]

$$M(t) = -P_0\sqrt{\frac{a}{b}}\sin(t\sqrt{ab}) + M_0\cos(t\sqrt{ab}),$$

$$P(t) = M_0\sqrt{\frac{b}{a}}\sin(t\sqrt{ab}) + P_0\cos(t\sqrt{ab}).$$

Here M_0 and P_0 represent Peach and Mario's initial feelings for one another (you can verify directly from the above equations that $M(0) = M_0$ and $P(0) = P_0$). Figure 7.9 shows the graphs of these equations look for different values of a, b, M_0 and P_0—the red graph is $M(t)$, the blue is $P(t)$.

This degree of mathematical sophistication isn't at work under the hood when you're playing a game like *The Sims*. But since friendship in *The Sims* isn't symmetric (just because Sim A likes Sim B doesn't mean the reverse is true), it's worth considering a model that also lacks symmetry. As games become increasingly realistic in other aspects of their design, it's only a matter of time before we see one that incorporates these ideas more explicitly.

7.8 IMPROVING THE MODEL (AGAIN!)

If you think about it, though, there's an obvious deficiency in our model. As it stands, the only types of relationships we can model are ones with

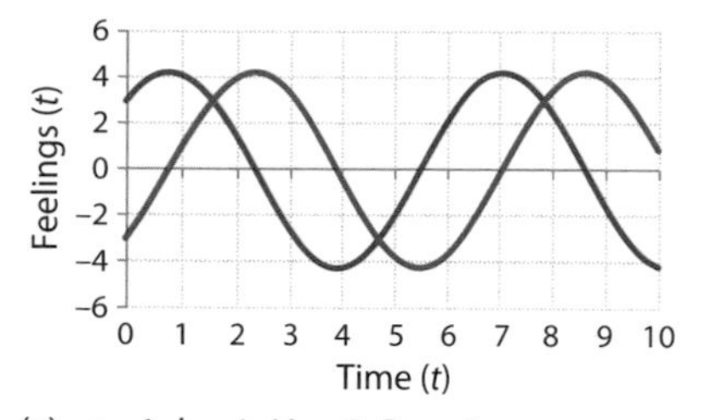

(a) $a = 1, b = 1, M_0 = 3, P_0 = -3.$
In this case, each partner's feelings vary between the same maximum and minimum, but their feelings rarely match.

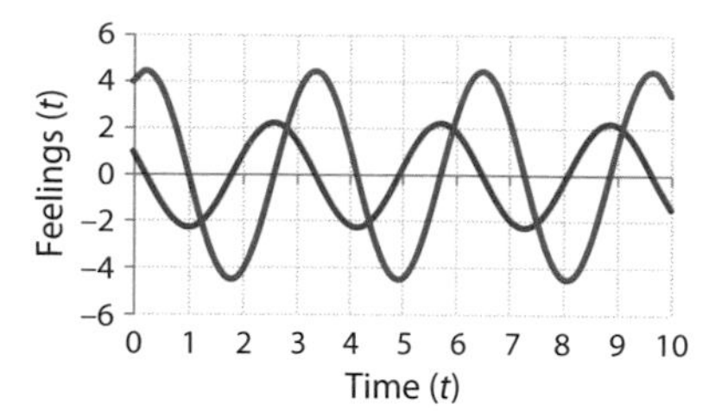

(b) $a = 1, b = 4, M_0 = 1, P_0 = -4.$
A larger value of b means Peach's feelings oscillate between higher highs and lower lows. Mario'se motions are somewhat less volatile.

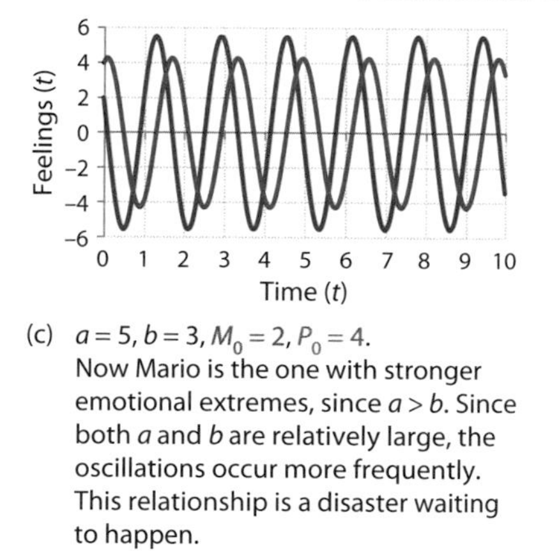

(c) $a = 5, b = 3, M_0 = 2, P_0 = 4.$
Now Mario is the one with stronger emotional extremes, since $a > b$. Since both a and b are relatively large, the oscillations occur more frequently. This relationship is a disaster waiting to happen.

FIGURE 7.9. Examples of different relationship dynamics for Mario and Peach.

perpetual peaks and valleys. But (fortunately!) not all relationships are so dramatic. Some peak early and peter out, some don't work from the outset, and the good ones stabilize at some relatively high value. So, the question becomes: How can we modify this model to try to capture a broader spectrum of relationships?

The first generalization (suggested in Strogatz's original note) is to allow Mario and Peach's feelings to change not only in response to the *other* person's feelings but in response to their *own* feelings as well. In other words, we can consider differential equations of the form

$$M'(t) = a \times M(t) + b \times P(t),$$
$$P'(t) = c \times M(t) + d \times P(t),$$

for some constants a, b, c, and d. Notice that if we set $a = d = 0$ and take b to be negative and c to be positive, we reduce to the case considered before. But as might be expected, this more general framework allows us to consider more general phenomena.[11] Notice that if all the coefficients are positive, the feelings Mario and Peach have for one another will grow exponentially and without bound. Conversely, if all the coefficients are negative, the two of them will quickly turn into bitter enemies, no matter the strength of their initial feelings for one another. If you're curious, you can explore what happens if some of the coefficients are positive and some are negative, and how the solutions change with the relative sizes of a, b, c, and d. Are Mario and Peach doomed to suffer through instability, or are there values for which their relationship stabilizes without exploding in one direction or another?

If you play around with the parameters of this model, you can discover some truly dysfunctional relationships. Long-term stable relationships exist as well, but as in the examples in Figure 7.10, "stability" almost always means that both partners' feelings for one another tend toward zero. Are we forced to conclude, then, that all relationships that don't explode are bound to peter out?[12]

Your answer may depend on how recently you've gone through a bad breakup. But for the sake of mathematical interesting-ness, let's take a less cynical, more analytic approach. In order to capture more nuanced behavior, though, we'll need to put even more information into the model.

One idea we can try to incorporate is that we're influenced not just by our feelings and our partner's feelings: we're also often attracted to a partner's intrinsic appeal. This idea was explored in a 1998 paper titled "Love Dynamics: The Case of Linear Couples," written by Sergio Rinaldi. In his model, the differential equations take on the following form:

$$M'(t) = -a \times M(t) + b \times P(t) + c \times P_{\text{appeal}},$$

$$P'(t) = d \times M(t) - e \times P(t) + f \times M_{\text{appeal}}.$$

Here each of the constants a through f is positive (in particular, it is assumed that one's feelings naturally decay over time, hence the

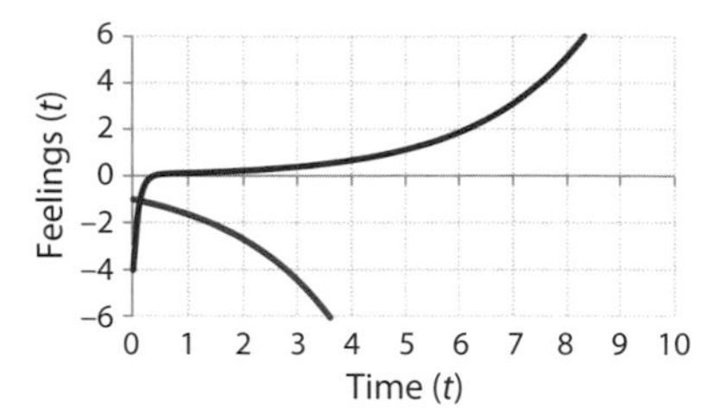

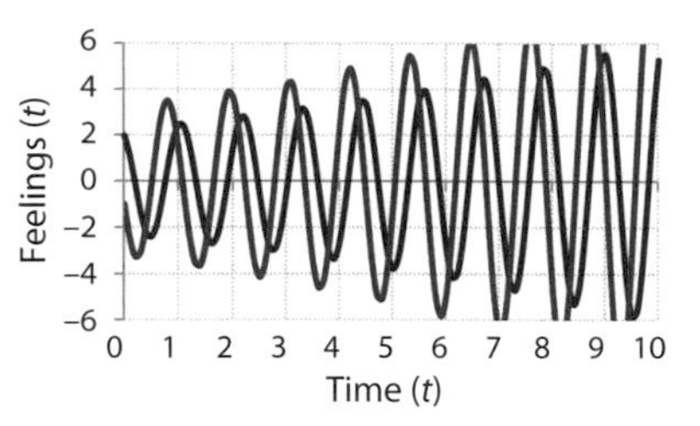

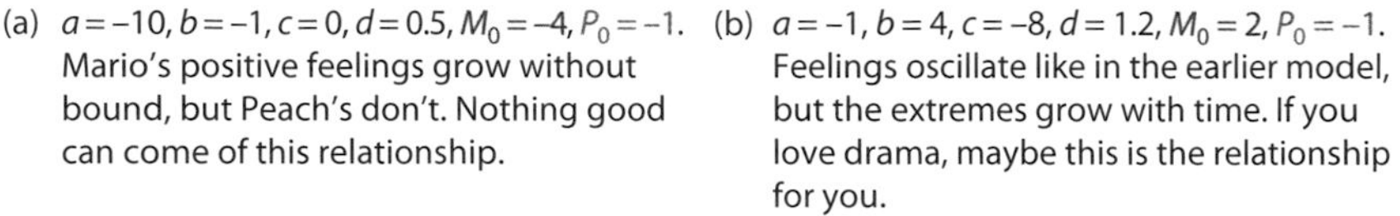

(a) $a = -10, b = -1, c = 0, d = 0.5, M_0 = -4, P_0 = -1$. Mario's positive feelings grow without bound, but Peach's don't. Nothing good can come of this relationship.

(b) $a = -1, b = 4, c = -8, d = 1.2, M_0 = 2, P_0 = -1$. Feelings oscillate like in the earlier model, but the extremes grow with time. If you love drama, maybe this is the relationship for you.

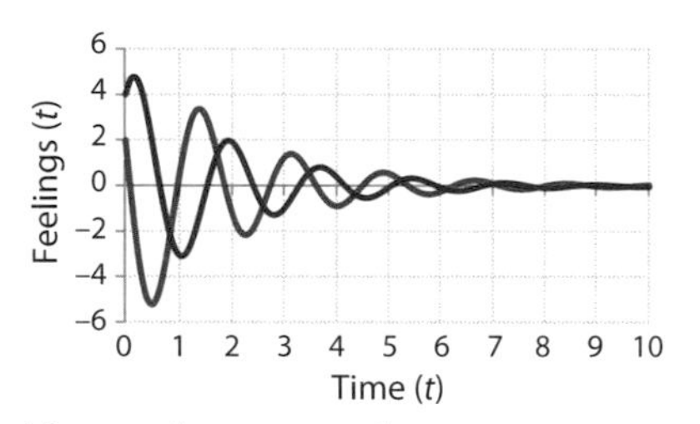

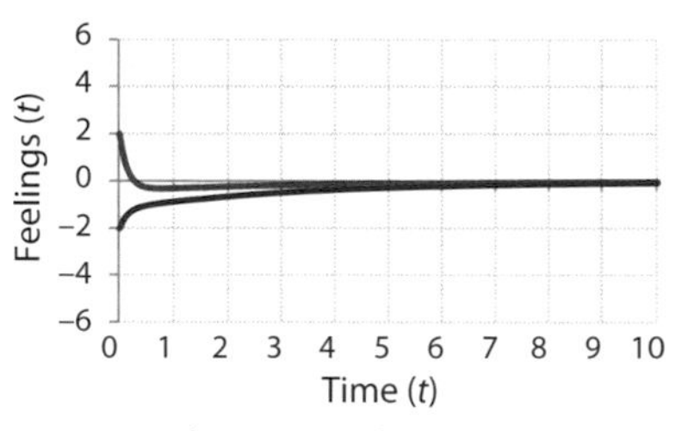

(c) $a = 1, b = 3, c = -5, d = -2, M_0 = 4, P_0 = -2$. Feelings stabilize in this case. Unfortunately, stability here means that each partner feels indifferent towards the other.

(d) $a = -1, b = 2, c = 2, d = -6, M_0 = -2, P_0 = 2$. This relationship is fairly stable too, but it's also the least interesting of the bunch.

FIGURE 7.10. Examples of dynamics in a more general model.

presence of the two negative signs). The constants c and f indicate how receptive each individual is to the appeal of the other person. For simplicity, each person's appeal is assumed to be constant.

This model has some interesting properties not shared by the other ones. First, provided that $b \times d < a \times e$, $M(t)$ and $P(t)$ will always stabilize—no more cyclic behavior or unstoppable exponential growth. What's more, provided each person's feelings are initially nonnegative, they will never be enemies. Instead, their feelings will strictly increase and approach their asymptotic values.

In fact, there are relatively simple formulas for each person's long-term feelings. According to this model, when Mario and Peach's feelings stabilize (that is, when $ae - bd > 0$), their stable values are given by

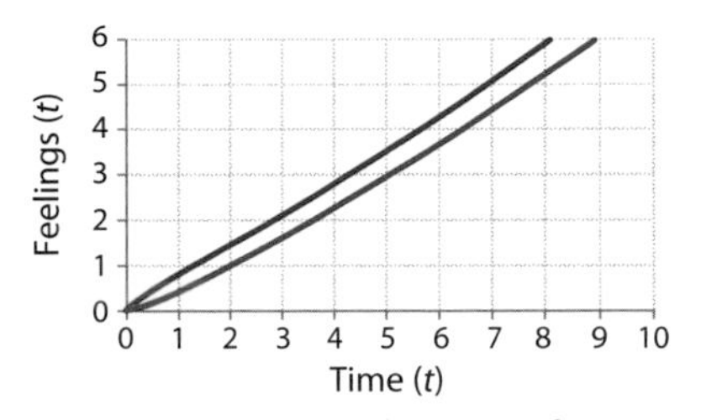

(a) $a = 1, b = 1.1, c = 1, d = 1, e = 1, f = 1.5$, $M_{\text{appeal}} = 0.1, P_{\text{appeal}} = 1$.
When $bd \geq ae$, this new model isn't much of an improvement. In this case, each person's feelings grow without bound, which doesn't seem realistic.

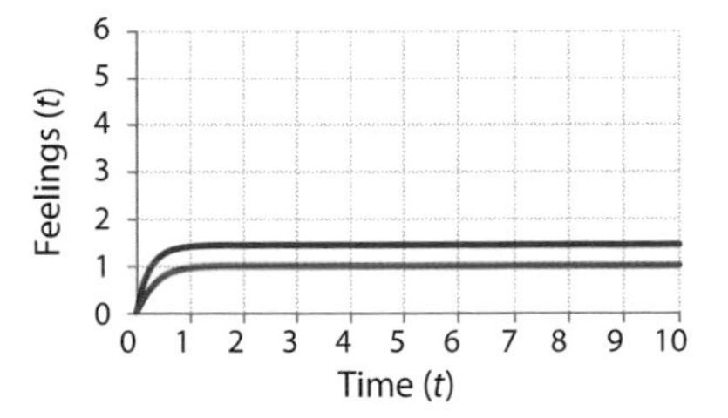

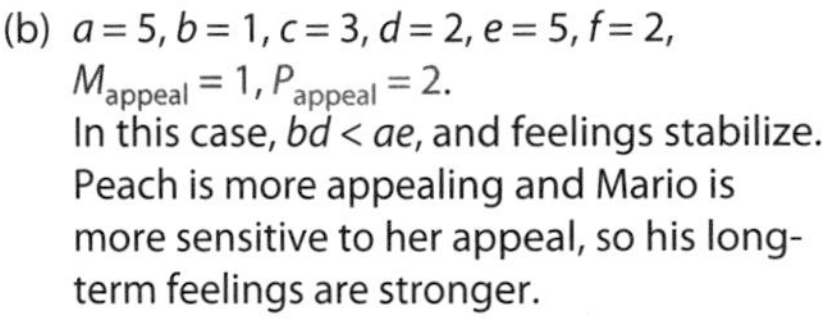

(b) $a = 5, b = 1, c = 3, d = 2, e = 5, f = 2$, $M_{\text{appeal}} = 1, P_{\text{appeal}} = 2$.
In this case, $bd < ae$, and feelings stabilize. Peach is more appealing and Mario is more sensitive to her appeal, so his long-term feelings are stronger.

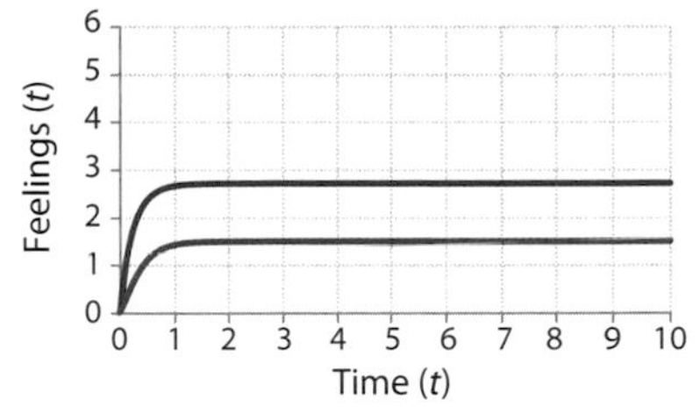

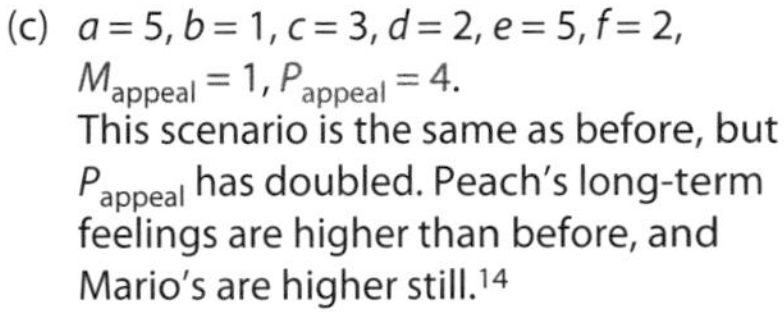

(c) $a = 5, b = 1, c = 3, d = 2, e = 5, f = 2$, $M_{\text{appeal}} = 1, P_{\text{appeal}} = 4$.
This scenario is the same as before, but P_{appeal} has doubled. Peach's long-term feelings are higher than before, and Mario's are higher still.[14]

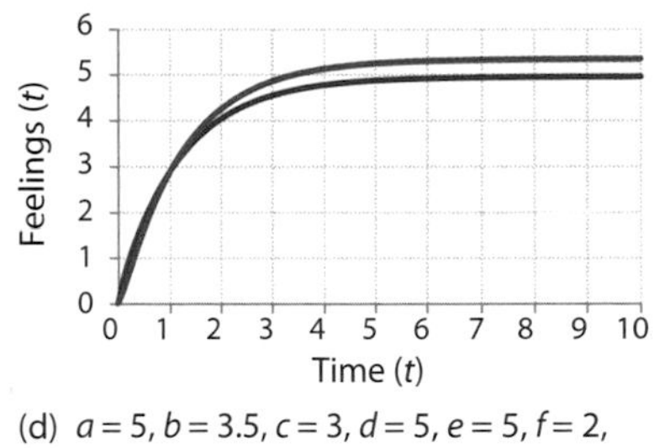

(d) $a = 5, b = 3.5, c = 3, d = 5, e = 5, f = 2$, $M_{\text{appeal}} = 1, P_{\text{appeal}} = 2$.
This closer $ae - bd$ is to zero, the higher the long-term feelings of each partner, and the slower the build towards those long-term feelings.

FIGURE 7.11. Examples of dynamics in an even more general model.

the following:

$$M_{\text{stable}} = \frac{ec\,P_{\text{appeal}} + bf\,M_{\text{appeal}}}{ae - bd},$$

$$P_{\text{stable}} = \frac{dc\,P_{\text{appeal}} + af\,M_{\text{appeal}}}{ae - bd}.$$

Note that, as with our earlier models, the long-term feeling scores don't depend on the initial values M_0 and P_0. So much for the importance of first impressions!

Figure 7.11 shows some examples of the types of behavior we can get with these revisions to our model. In each case, we've taken $M_0 = P_0 = 0$, which seems reasonable, since when people first meet they are probably indifferent to one another.[13]

These examples illustrate one of the more interesting conclusions of Rinaldi's paper: whenever feelings stabilize, those stable values are higher precisely when the romance builds more slowly. Put another way, you should watch out for those passionate affairs: the slower the burn, the higher the long-term satisfaction.

The fact that relationships can stabilize with this model is a promising sign, though by Rinaldi's own admission, it's still a simplified version of reality. Rinaldi has since published several more papers refining this model. And he's not the only one: by now, the field of mathematical sociology has considered a number of generalizations.

All of these cases are examples of *linear* systems of differential equations, and the solutions have mathematically predictable long-term behavior. But as you know, interpersonal relationships, like weather patterns and changes in the stock market, are highly unpredictable.

One proposed change to the model is the addition of time delays. The idea here is that our emotions don't necessarily respond to the moods of our partner instantaneously. Such a model might look something like this:

$$M'(t) = -a \times M(t) - A \times M(t - t_1) + b \times P(t)$$
$$+B \times P(t - t_2) + c \times P_{\text{appeal}},$$
$$P'(t) = d \times M(t) + D \times M(t - t_3) - e \times P(t)$$
$$-E \times P(t - t_4) + f \times M_{\text{appeal}}.$$

Here the parameters t_1, t_2, t_3, and t_4 represent delays; in this case, emotional states change in response to both current and past feelings.

People also add complexity into these models by introducing other *nonlinear* terms. In these models, the rates of change don't depend just on linear combinations of $M(t)$ and $P(t)$, but also on higher-order terms, like $M(t) \times P(t)$.[15] As you can imagine, the models can get quite complicated very quickly. There's a tradeoff here: more complex models may be more realistic, but if they're too complex they may defy general analysis, or they may be so specific that the conclusions they imply aren't widely applicable. As Einstein said,[16] "Everything should be made as simple as possible, but not simpler." Whether applied to the models describing love, or to love itself, the advice is equally sage.

7.9 CONCLUDING REMARKS

As anyone who's ever been involved in one knows, human relationships are complicated. It shouldn't be surprising, then, that the mathematics involved can quickly become quite sophisticated. Friendship models in games like *Far Cry 2* are clearly not representative of reality, and while simulations like *The Sims* get a little closer, there's still room for improvement. As games continue to grow in complexity, it would be nice to see their relationship models also align themselves more closely to some of the mathematical models presented here.

At the very least, I hope Murray will think a bit more carefully about how he tracks the changes in his friendships over time.

8
Order in Chaos

8.1 THE ESSENCE OF CHAOS

One of the most famous portrayals of a mathematician in popular culture comes to us courtesy of *Jurassic Park*. In the 1993 blockbuster, Jeff Goldblum plays Ian Malcolm, a fictional mathematician who helped popularize the term "chaos theory."

In the film, Dr. Malcolm describes his field of study as follows:

> It simply deals with unpredictability in complex systems. The shorthand is the butterfly effect: a butterfly can flap its wings in Peking and in Central Park you get rain instead of sunshine.

This isn't the only time the *butterfly effect* has made an appearance in film. The phenomenon even took on a starring role in the 2004 film *The Butterfly Effect*, in which Ashton Kutcher travels back in time only to find that small changes in the past can have a profound effect on the present.

For all the buzz generated by terms like "chaos theory" and "butterfly effect," though, the mathematics that inspires these terms is often overlooked. This is a shame, because the underlying mathematics is much more interesting than the terms used to describe it. Chaotic systems are also fairly common, both in the world around us and in the games we play. In fact, we've encountered a few examples of them already, though we haven't (yet!) gone into much detail about them.

In this chapter, we'll dig into the mathematics of dynamical systems a little bit deeper. We'll explore what makes a system chaotic, see some examples of chaotic systems inspired by games, and talk about the difference between a random system and a chaotic one. There won't be any dinosaurs, unfortunately, but I'm confident that we can give Dr. Malcolm a run for his money.

8.2 LOVE IN THE TIME OF CHAOS

In the previous chapter, we explored a few classes of differential equations used to model relationship dynamics. The models we considered in the most detail all highlighted *linear* differential equations. And although we gave some examples of *nonlinear* systems of differential equations, we didn't delve into too much detail.

Nonlinear systems are qualitatively different from linear ones, and though they're commonly found in nature, they're also much harder to deal with. One reason for this is that it's harder to generate solutions. For example, if you have a linear differential equation and you know that f and g are two functions that solve the system, you also know that $af + bg$ will solve the system for any numbers a and b. But in general, it's not possible to combine solutions to nonlinear systems in this way in order to generate new solutions.

More importantly for our current discussion, nonlinear systems can exhibit chaotic behavior, while linear systems do not. Let's explore what this means within the context of another example.

In 2004, Professor J. C. Sprott of the University of Wisconsin, Madison, wrote an article titled "Dynamical Models of Love," which incorporated nonlinear terms into a model we've previously explored:

$$M' = a \times M + b \times P,$$
$$P' = c \times M + d \times P.$$

(Remember that we're considering two people, whom we named Mario and Peach. Mario's feelings are given by M, while Peach's are given by P.[1])

Sprott introduces nonlinear terms into the model by replacing these equations by the following ones:

$$M' = a \times M + b \times P\,(1 - \lceil P \rceil),$$
$$P' = c \times M\,(1 - \lceil M \rceil) + d \times P.$$

Here, the function $f(x) = \lceil x \rceil$ is referred to as the *ceiling* function, which rounds any number up to the nearest whole number. For example, $\lceil 2.5 \rceil = 3$, $\lceil 4.00001 \rceil = 5$, and $\lceil 10 \rceil = 10$. Since the right-hand sides of

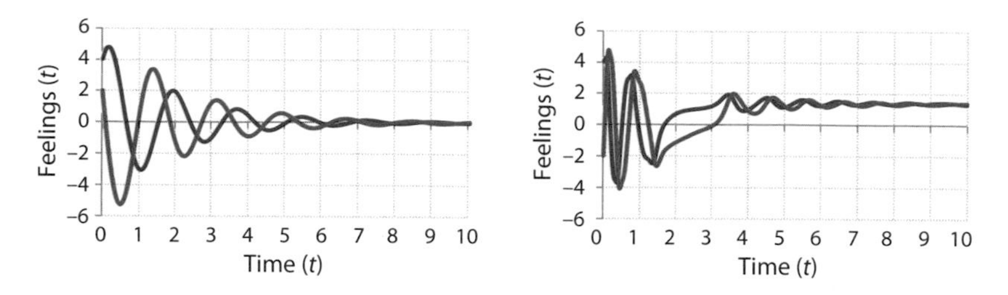

FIGURE 8.1. The graph on the left shows M and P under the linear model; the right uses the nonlinear model.

the above equations now involve more than linear combinations of M and P, these equations are considered nonlinear. Specifically, the terms on the right-hand side of each equation involving the ceiling function are called *nonlinear* terms.

To understand what this modification means, let's just look at the first equation. For example, if $b > 0$, then originally Mario would respond positively to Peach's feelings, no matter how strong. In reality, though, if Peach's feelings were too strong, Mario would likely start to feel smothered and retreat. Conversely, if Peach was really angry with Mario and he wanted to salvage the relationship, he might respond positively to her feelings and try to repair whatever damage he caused. This changing response to Peach's feelings isn't captured in the first model, but is captured by the addition of the $(1 - \lceil P \rceil)$ term. In this model, when P moves from being less than 1 to being greater than 1, for instance, Mario's response to Peach's feelings qualitatively changes.

If we look back at some of the scenarios from the previous chapter, we'll find that this new model gives some different results. For example, in the case where $a = 1$, $b = 3$, $M_0 = 4$, $c = -5$, $d = -2$, and $P_0 = -2$, the graph on the left-hand side of Figure 8.1 shows M and P under our earlier model, while the graph on the right-hand side shows M and P under our new model. Notice that unlike our original linear model, this nonlinear model allows for solutions with steady states away from zero.

This is all interesting, but it's hardly chaotic. In order to get some chaotic behavior, much like in the real world, we need to add a third party. Imagine now that Peach has a second suitor; call him Wario. Mario and Wario don't know about each other, but both know about

Peach. Peach's feelings respond to both Mario and Wario, but the feelings of both men respond only to Peach. By generalizing the nonlinear equations from above, we can come up with a slightly more complicated system, one that's also featured in Sprott's paper. To do this, we need to separate out Peach's feelings into two pieces: P_M represents her feelings toward Mario, and P_W represents her feelings toward Wario. This gives us four equations in total (M, W, P_M, and P_W), whose rates of change are given by the following:

$$M' = a \times M + b \times P_M (1 - \lceil P_M \rceil),$$

$$P'_M = c \times (M - W)(1 - \lceil M - W \rceil) + d \times P_M,$$

$$W' = e \times W + f \times P_W (1 - \lceil P_W \rceil),$$

$$P'_W = c \times (W - M)(1 - \lceil W - M \rceil) + d \times P_W.$$

If you feel like you're lost in a sea of equations, don't worry. The equations are less important for us than the pictures that they describe. As before, depending on the values of a, b, c, d, e, and f, you can get all sorts of unrealistic behavior: feelings that grow exponentially without bound, or feelings that oscillate wildly. But this new setup also provides us with examples of a new class of relationship.

Here's an example: suppose that $a = -4$ and $b = 1.1$, so that Mario distrusts his own feelings but trusts Peach's (up to a point). Let's assume that Peach is the same way, with $c = 3$ and $d = -2$. Wario, however, trusts his own feelings, and for him, $e = 1.5$ and $f = -1$. Let's also assume that Mario and Peach both start with positive feelings for each other ($M_0 = P_{M0} = 1$), but that Wario and Peach are indifferent.

Given all this starting information, Figure 8.2 shows how these two romances will play out over time. If prompted, how would you describe the graphs? They're not periodic, nor do they appear to be growing or shrinking without bound. Each person's feelings ebb and flow, but not in a way that's easy to describe or predict. Perhaps the best thing we can say about the parties involved is that they seem, well, temperamental.

But the great thing is that they're temperamental in a mathematically meaningful way. To understand what I mean by this, let's crunch the numbers once more. We won't change any of our original parameters,

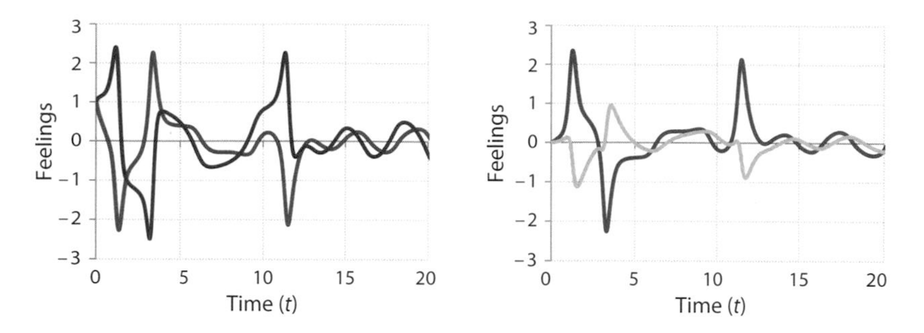

FIGURE 8.2. Mario's and Peach's feelings for one another are on the left; Wario's and Peach's are on the right.

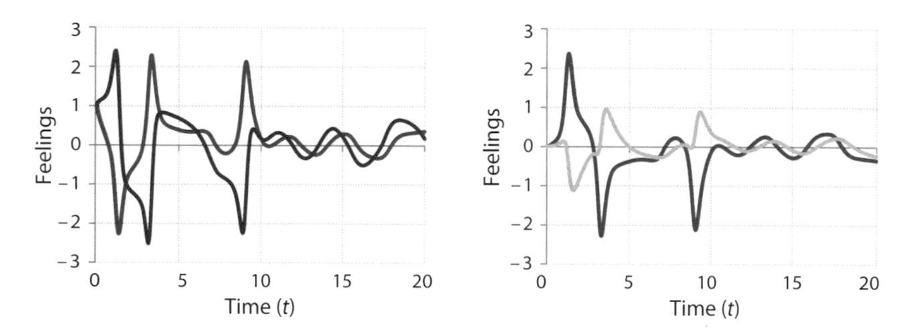

FIGURE 8.3. The same relationships as before, with slightly different initial conditions. Note that even though Wario's and Peach's initial feelings are unchanged, their relationship evolves differently. This is because Peach's feelings depend on Mario's in both relationships.

except for one: let's assume that Mario feels just a *bit* more strongly than we previously assumed, and let's increase M_0 from 1 to 1.01.

This may not seem like much, but this small change results in some startlingly different graphs. The new graphs are shown in Figure 8.3. Also, for ease of comparison, Figure 8.4 shows graphs of Mario's and Peach's feelings for one another in both scenarios. Mario's are on the left, Peach's are on the right.

These figures illustrate a simple fact: even a small change in one person's initial feelings can have a drastic effect on how those feelings evolve over time. The butterfly flaps its wings, and chaos emerges.

Of course, deviation on its own isn't enough to suggest chaos. After all, the curves we made before Wario entered the scene will also shift if you change the initial parameters (Figure 8.5). What makes the

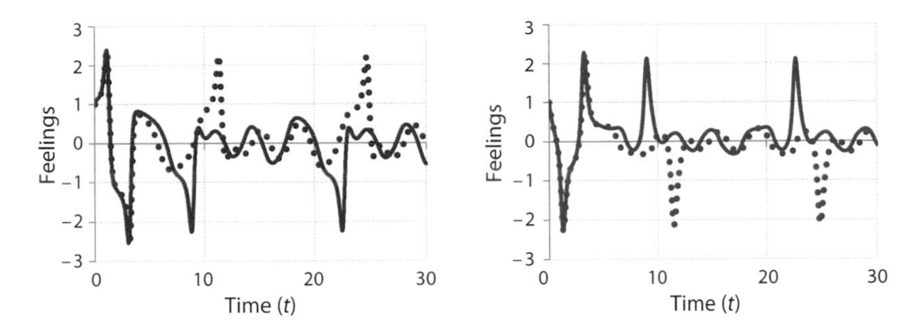

FIGURE 8.4. The dashed curves correspond to the case $M_0 = 1$; the solid curves correspond to $M_0 = 1.01$.

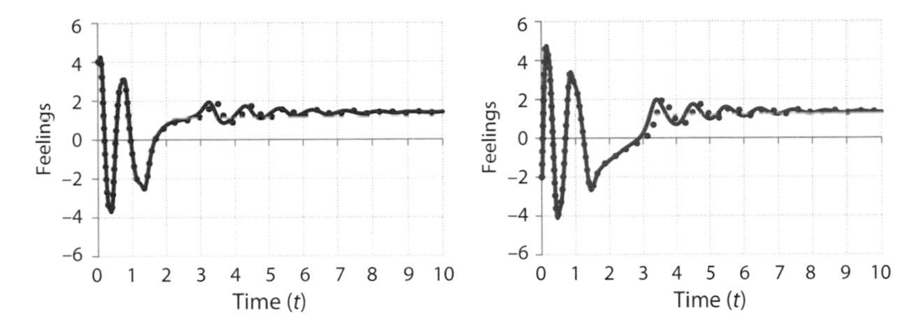

FIGURE 8.5. Solutions to the earlier two-person nonlinear system when $a = 1$, $b = 3$, $c = -5$, $d = -2$, $P_0 = -2$, and $M_0 = 4$ (dotted line) or 4.01 (solid line). In this case, there's much less sensitivity to initial conditions.

dynamics chaotic isn't the existence of deviation when you vary the initial conditions. Rather, it's the *degree* of deviation: for a chaotic system, trajectories that begin close to one another will deviate at an *exponential* rate. This means that making long-term predictions is next to impossible, since even small errors in initial measurements can have a huge effect down the road and can't be accounted for in any meaningful way. In mathematical parlance, this property is called *sensitivity to initial conditions*, and it is a hallmark of chaos.[2]

For some of us, when it comes to human relationships, chaos may sound like the norm. It's certainly not the case that we can accurately predict the strength of all our current relationships a decade from now. But I hope that there are a few relationships in your life that are relatively stable. Too much chaos can be hard to take.

FIGURE 8.6. Trajectories from firing a shell perfectly horizontally (left), vertically (middle), and at a 45° angle (right).

8.3 SHELL GAMES REVISITED

When we explored pursuit and evasion in Chapter 5, we spent some time analyzing the trajectories of green and red shells in *Mario Kart* games. You may remember that green shells are perpetual motion machines and bounce around a course until they hit a target. Red shells, on the other hand, actively seek out your opponents, but are less durable: if one hits a wall, it will break apart instead of bouncing back.

We analyzed a few potential green-shell trajectories in the simple case where you're on a straight track and one of your opponents is directly in front of you. We didn't take our analysis further for practical reasons: namely, the trajectory of a shell quickly becomes difficult to describe, especially if the shell never hits anything.

Even so, there are some interesting observations we can make about different *classes* of green-shell trajectories. And, as with relationship dynamics, in the right circumstances it's possible for chaos to emerge.

Before diving into the deep end, let's first consider a simplified course, one that's square and contains no obstacles. A shell fired in this course bounces off the walls of the square in perpetuity until it hits a target.

Imagine you fire a shell from the center of the square. In certain cases, the trajectory is fairly easy to describe mathematically. For example, if you fire it horizontally or vertically, the shell will just bounce from left to right or from top to bottom. Firing the shell at a 45° angle, so that the initial slope of the trajectory is 1, is similarly uninteresting (Figure 8.6).

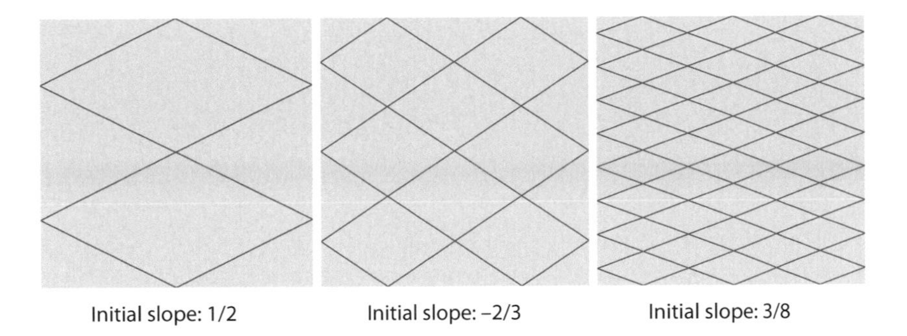

FIGURE 8.7. Examples of periodic trajectories starting from the center.

By varying the trajectory, there are an infinite number of paths we could examine. Figure 8.7 highlights a few more, classified by the initial slope of the trajectory.

You may notice that in each case, the trajectory of the shell is *periodic*: the shell departs from the center of the square, bounces off some walls, and then returns to its initial state. For instance, in the case where the shell has an initial slope of 1/2—that is, where it starts by moving two pixels to the right for every pixel it moves up—the shell will trace out a sort of figure-eight pattern on repeat. The other initial slopes lead to more intricate patterns, but in both cases they are *repeating* patterns.

These repeating patterns could be exploited if you knew how to spot them. For instance, if you knew you were being chased by a green shell with the figure-eight pattern seen in Figure 8.7, you could simply move to one of the corners of the board: the shell would never come near you, and you could formulate a plan of attack in relative safety.

Unfortunately (or fortunately, depending on your perspective), not all trajectories lead to periodic orbits. For example, Figure 8.8 shows snapshots of a trajectory over time for a shell with an initial slope of $\sqrt{2}$.

Of course, on their own these diagrams may not be totally convincing. After all, isn't it possible that this shell has a periodic trajectory but that its period is simply too long to observe on a short time scale?

It's a reasonable thought, but it's also incorrect. The truth is that whenever the slope of the trajectory is *irrational*—that is, whenever the slope can't be expressed as a fraction p/q for integer p and q—the trajectory of the shell will not be periodic. Since $\sqrt{2}$ is an irrational

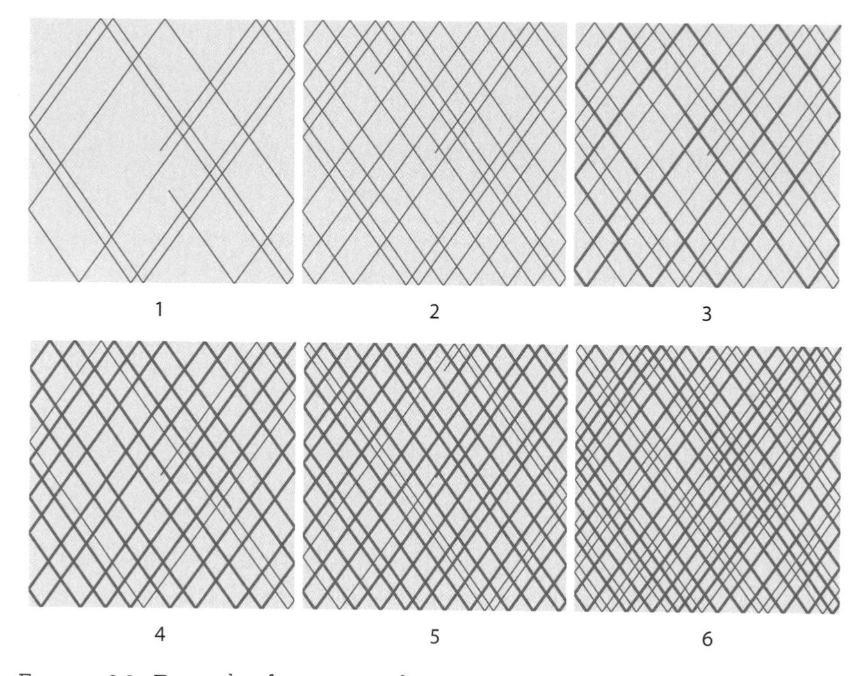

FIGURE 8.8. Example of a nonperiodic trajectory over time (initial slope is $\sqrt{2}$).

number, the trajectory pictured in Figure 8.8 won't ever repeat itself.[3] What's more, when the trajectory of a shell has an irrational slope, the shell will eventually reach every point on the board (though it will never touch the same spot twice).[4] Put another way, the trajectory will eventually fill up the entire board, and it will do so in a uniform way; that is, it won't favor one area of the board over another as it's bouncing around.[5] If you pick any two regions of the course, the shell will spend, on average, an equal amount of time in both: this property is known as *ergodicity*.

In terms of your *Mario Kart* strategy, ergodic trajectories are bad news for a few different reasons. First, as we've already seen, these trajectories fill up the entirety of the board: staying in one place may not be such a smart strategy anymore. Secondly, even if you knew that the shell's trajectory was ergodic, making predictions about the future would be difficult unless you could measure the slope of the trajectory perfectly. For example, Figure 8.9 shows images of two trajectories over time: the green path corresponds to our old friend $\sqrt{2} = 1.4142\ldots$, while the blue path corresponds to $\sqrt{2.01} = 1.4177\ldots$

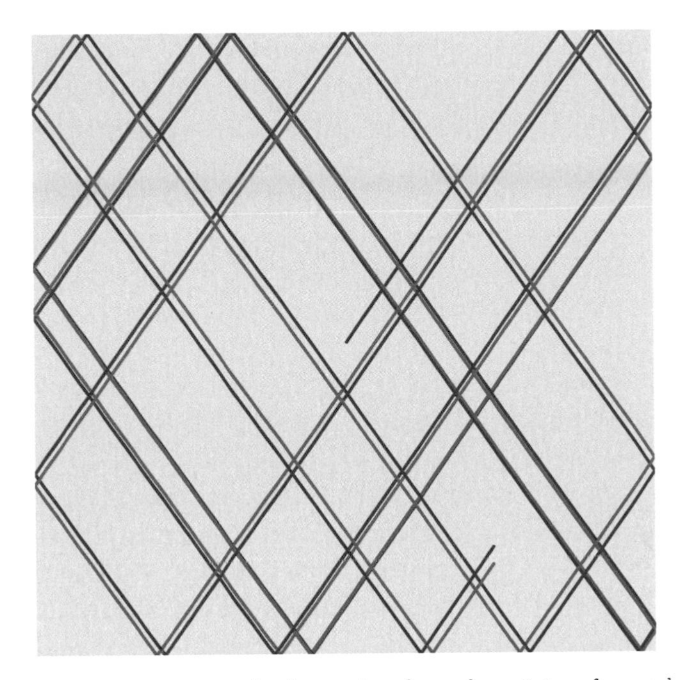

FIGURE 8.9. Two trajectories, both starting from the origin, after 20 bounces. As you can see, the blue trajectory is gradually moving away from the green one.

Even though the slope of the blue trajectory differs from the slope of the green trajectory by less than 0.25%, over time the position of these two shells will deviate more and more. Having said that, the rate at which they deviate is linear, not exponential. Ergodicity alone does not chaos make.

All of this might not be so bad, though, if there were relatively few ergodic trajectories. But this brings us to our third dilemma. As we've seen, ergodic trajectories correspond to irrational numbers, while periodic trajectories correspond to rational numbers. So when we want to compare the number of ergodic trajectories to the number of periodic ones, really what we want is to compare the number of irrational numbers to the number of rational ones. And here we come to another wrinkle: while there are infinitely many of each type of number, there are, in some sense, infinitely *more* irrational numbers.

Typically the comparison is framed like this: the rational numbers are what's called a *countable* set. They're countable because, if you wanted to, you could enumerate the rational numbers into an (infinitely long) list.[6] This is not true, however, for irrational numbers. If you've ever tried to enumerate all the *real* numbers—that is, the collection of both the rationals and the irrationals—you'll soon find that you run into a problem.

To see why, let's focus our attention on just the numbers between 0 and 1. Let's say you tried to come up with an enumeration of all such numbers, based on their decimal expansion.[7] Your list might start like this: 0.809565, 0.308752, 0.253827, 0.957324, 0.777709, ...

Suppose, once you're done with this infinitely long list, you pass it along to someone to verify that you have indeed included every number. If your reviewer has her wits about her, she'll likely notice that you've omitted something.

For instance, your reviewer could present you with a decimal number where the nth digit after the decimal is 0, unless the nth digit in the nth number on your list is a 0, in which case the nth digit is 1. Based on the list above, this "new" number would begin as follows: 0.01001.... And indeed, this number wouldn't match up with any of the numbers in your list, since it would differ from the nth number at the nth digit. This argument works for *any* list of real numbers, and therefore *any* attempt at enumerating the real numbers will fail. But the issue can't come from the rational numbers, which *can* be put into a list; therefore, the irrationals must be the culprit.[8] So, in this idealized *Mario Kart* model, almost all trajectories are ergodic.

Of course, all of this analysis rests on the assumption of a very simple course: one that is square and has no obstacles. When we talked about *Mario Kart* before, we focused primarily on the Block Fort course, which is square, but with smaller square obstacles inside it. Does this change our analysis at all?

Not really. In fact, in many cases the results in this simple example—namely, that almost all[9] trajectories are ergodic—still hold. In order for these results to still hold, the salient features of the course are the following:

1. The course must be a *polygon* (in this case, a square).
2. The interior angles of the course, when measured in radians, must all be *rational* multiples of π (in this case, all of the angles are $\pi/2$).

FIGURE 8.10. Both sample trajectories start below the bottom-left corner of the red block, half-way between the corner and the bottom edge of the board. The left trajectory is periodic and has an initial slope of 1. The right trajectory has an initial slope of $\sqrt{3}$.

3. All obstacles on the course must also be polygons whose interior angles are rational multiples of π.

Whenever these three conditions hold, almost all trajectories are ergodic and will fill up the entire board, rather than tracing out some periodic path.[10] Trajectories in such regions are usually contextualized using the game of billiards, which is why the phrase *rational polygonal billiards* is typically what's used to describe this type of phenomenon.[11]

And indeed, the Block Fort course in *Mario Kart 64* fits all of these conditions.[12] Consequently, almost all green-shell trajectories are ergodic, and in the heat of battle, very difficult to trace over any significant time scale. A couple of examples are shown in Figure 8.10.

Ergodicity is nice, but can we find trajectories that are truly chaotic? In the case of rational polygonal billiards, the answer is no. But what about boards that aren't rational polygons containing rational polygonal obstacles? Well, in this case, all bets are off. For example, on elliptical boards, it's possible for a trajectory to be neither periodic nor ergodic (Figure 8.11).[13]

And indeed, other battle courses in *Mario Kart* don't fit into the rational polygonal mold. One of the other battle courses in *Mario Kart 64*, for instance, is called Big Donut. As you might expect, the course

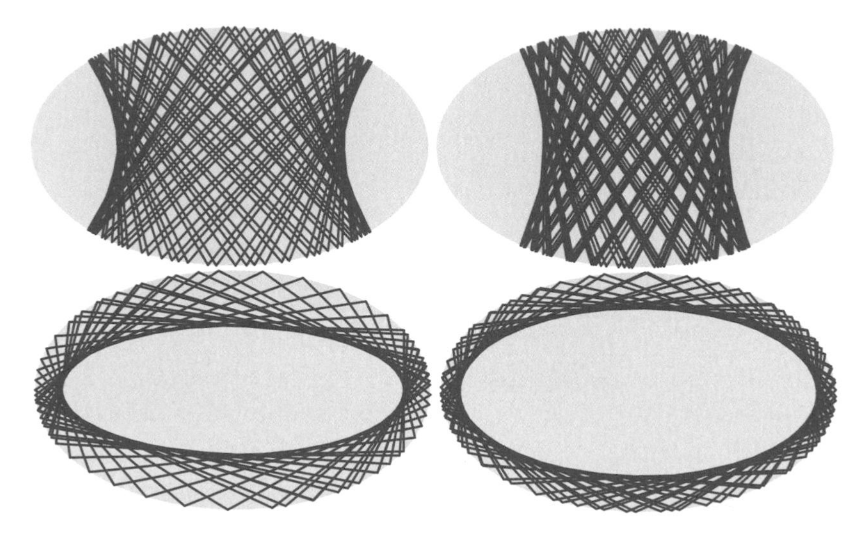

FIGURE 8.11. Sample trajectories on an ellipse, with varying initial positions and slopes.

looks like a doughnut—it's a circular course, with a hole cut out in the middle. Unlike a doughnut, though, in that center is a pit of hot lava. Yet even with a course that has an instant death trap in the middle, it should be possible to fire a shell that will travel along the course in perpetuity. Can you think of a trajectory that might work?

While both of these types of courses are interesting, neither one exhibits chaos. All is not lost, though. In fact, by combining the straight edges from our first example with the rounded edges of our second, chaos finally emerges.

One of the more classical examples of billiard table that exhibits chaos is called the "Bunimovich stadium" and consists of a rectangle with semicircles on either end. A course like this does exhibit sensitivity to initial conditions, as demonstrated in Figure 8.12.

In other words, when it comes to *Mario Kart*, the existence of true chaos depends not on how you fire your shell but on the course you've selected for battle.

In its review of the Wii U game *Mario Kart 8*, Mashable.com noted that one of the best aspects of the game was its "chaotic, fast-paced multiplayer."[14] While the gameplay may not always technically be chaotic, this statement is, on occasion, mathematically precise. And even when

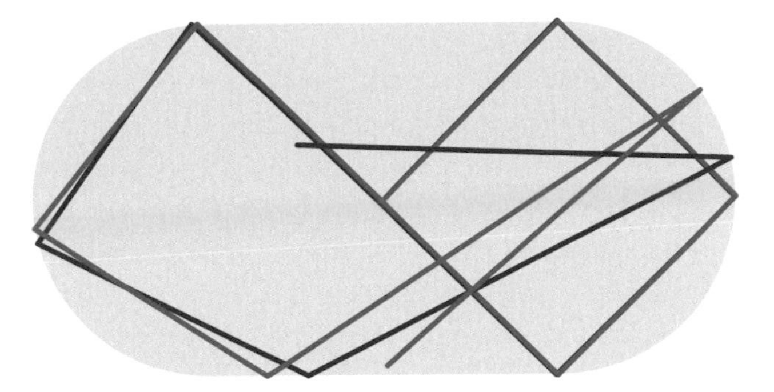

FIGURE 8.12. Two trajectories are shown, both starting at the origin. The green trajectory has an initial slope of 1, and the blue has an initial slope of 1.01. In this example, it takes only seven bounces before the two trajectories look like they have nothing to do with one another.

the gameplay itself isn't chaotic, the relationship dynamics among four friends playing the game surely must be. Rarely are emotions as volatile as during a heated *Mario Kart* match.

8.4 HOW'S THE WEATHER?

Before moving on from these examples of chaos, let's take a quick look at one common misconception. Namely, chaotic behavior is *not* the same as random behavior. Certainly chaos and randomness can look similar; if you told people that the feeling graphs for the love triangle among Mario, Wario, and Peach were generated by some random process, they would have little reason to doubt you.

But for chaotic systems, any apparent randomness is just an illusion. In fact, chaotic systems are the exact opposite of random: they are completely *deterministic*. For the same set of inputs, a chaotic system, without fail, spits out the same output. While the output may be *unpredictable*, it is no more random than the output when you ask your calculator to evaluate $1 + 1$.

Because of this apparent randomness, though, chaotic processes are often modeled by random processes. While random modeling may not always be the most accurate reflection of reality, the outcome is often just as convincing.

One example of this interplay between randomness and chaos within video games can be found in weather modeling and prediction. The weather is probably the quintessential example of a chaotic system: indeed, it is the system at the center of the butterfly effect analogy. Equations used to model the dynamics of the atmosphere are, like the relationship equations we saw earlier, nonlinear in general. They can result in chaos in even simple cases; in fact, meteorologist-turned-mathematician Edward Lorenz published a 1963 paper showing that even an extremely simplified model for atmospheric convection, consisting of just three equations, exhibits chaos.[15] And despite advances in meteorology over the past fifty years, we're still unable to make reliable forecasts more than a week out. One of the main reasons for this is chaos. Even if we had perfect knowledge of the equations governing the dynamics of the atmosphere, we still wouldn't be able to measure quantities like temperature and pressure to an arbitrary degree of precision. As a result, sensitivity to initial conditions means that our predictions become less reliable the farther out we look.

It's no wonder, then, that modeling atmospheric dynamics in a video game is a less involved affair. Modeling weather effects based on our most realistic models is probably not the best use of resources (both in hours and computation time). Moreover, it's not necessary: a random process can be just as effective in modeling weather as a chaotic dynamical system.[16]

For instance, the open-world game *Skyrim* features fairly robust weather modeling. Some days are sunny and clear; others are cloudy and foggy. There can be light rain, severe thunderstorms, or even snow. But none of this comes from a truly dynamic model of the game world's atmosphere. Instead, each weather type is assigned some probability of occurring within a given time period. In fact, some *Skyrim* players have modified the game to add even more weather options, though the underlying probabilistic model isn't changed.[17] Still others have modified the game simply to make some types of weather much more likely than they were originally designed to be. So if you've been looking for that *Skyrim*/Pacific Northwest mashup, fear not: you can find a mod that will make it rain with much greater frequency.

In a way, this sort of probabilistic modeling is flipping meteorology on its head: rather than the probabilities being the *outputs* of the model, in games they're more often the *inputs*. In the real world, when a meteorologist says there's a 20% chance of rain, that's a conclusion based on an analysis of our model of the atmosphere. But in a game like *Skyrim*, these probabilities are the building blocks of the weather model, not an outcome of it.

The distinction may not matter much when you're in the heat of battle. But tempting though it may be, it's important not to conflate randomness with chaos. A random process and a chaotic process may look the same, but a chaotic process is reproducible given the same inputs; a random process is not.

8.5 CONCLUDING REMARKS

In 2010, Nintendo released *Donkey Kong Country Returns* for the Wii. The game is a sequel to the popular SNES *Donkey Kong Country* series from the 1990s, and it reunites Donkey Kong with his pal Diddy Kong (no relation to the hip-hop mogul) as they fight enemies and collect bananas.

The *Donkey Kong Country* series has always been heavy on platforming, and this game was no exception. Donkey and Diddy jump across chasms, swing on vines, and shoot themselves out of barrels. But don't let the bright colors and cartoon aesthetic fool you: in her review of the Nintendo DS version of *Donkey Kong Country Returns* for gaming site Polygon, Danielle Riendeau wrote that the game was "considerably harder than the SNES games" in the series.

Perhaps anticipating this complaint, the game came with a feature called the "Super Guide."[18] If you died on a level eight times in a row, you were given the option to essentially let the game play itself. When instructed by the player, Donkey Kong would move on autopilot, running and jumping through the level until the player decided to tap back in. In principle, this meant that the developers didn't have to choose between making the game accessible and making it challenging for long-time fans of the series.

In practice, however, implementing the Super Guide proved to be more difficult than Retro Studios, the development house behind the game, initially anticipated. In a postmortem of *Donkey Kong Country*

Returns at the 2011 Game Developers Conference, it was revealed that the developers "had to rewrite the engine to make it deterministic—make it that all the inputs would result in the same output every time, essentially eliminating randomness."[19]

In fact, the game likely would have needed to eliminate any chaotic dynamics as well, since the player has choice about when and where he wants the game to take control. In other words, the output—that is, the game's control of itself—must not be sensitive to initial conditions. Otherwise, starting the Super Guide a pixel to the right could mean the difference between life and death.

Retro was able to make the game predictable enough that it could play itself, but still hard enough to pose a challenge to players.[20] Even so, it's telling that when the game's sequel, *Donkey Kong Country: Tropical Freeze*, was released in 2014, the Super Guide was nowhere to be found.

The real world isn't designed to be lived on autopilot. So why should our games be? Randomness and chaos aren't necessarily bad things: in the right amounts, they can enrich and deepen the games we play, the relationships we have, and the lives we live. It shouldn't be surprising, then, that they lead to some beautiful mathematics too.

9
The Value of Games

9.1 MORE IMPORTANT THAN MATH

Several years ago, my now-wife and I took a trip to Denver to visit some close friends. One day, we went to the Denver Art Museum. We spent most of the afternoon taking in the exhibits, which ranged from Native American wood carving to contemporary art.

In some of the exhibits, the museum provided communal notebooks. I'd never seen this before, but the idea is fairly simple: the notebooks are there so that people can write down their thoughts about the exhibit as they walk through. It's a sort of analog commenting feature that serves as one way to start a conversation about the art on display. Unlike commenting boards on the Internet, though, the comments in these notebooks tended to be relatively thoughtful, positive, and free of profanity.

As I perused these public logs of other peoples' thoughts, I came upon a note written by a young student who had clearly been moved by the art on display. She had even written a poem in the notebook about the importance of keeping art programs in schools. The main thrust of her piece was that studying art was significantly more important, and more valuable, than studying subjects like math or science. I don't remember the entirety of what she wrote, but I do remember that she closed her poem with the following line:

Ideas are more important than math.

As you might expect, this line offended me on a visceral level. *This doesn't make any sense!* I thought to myself. *Math is nothing* ***but*** *ideas!* And not just any ideas, either—through mathematics, people have developed some of the most beautiful ideas that you'll find anywhere, in any medium.

TABLE 9.1. Instruction at Two Similar Schools

	Phoenix Park	Amber Hill
Socioeconomic Status	Less affluent	More affluent
Classroom Pedagogy	Less traditional	More traditional

At first, in spite of my better judgment, I remember being deeply upset at this student. To help me cope, I was already painting an unflattering picture of her in my mind. Of course, upon further reflection, I recognized that the fault was not hers. She was merely a reflection of the problems that so many students face in a traditional math classroom. Imagine if we taught music the way we traditionally teach math: we would give students pages of musical notation, explain to them how to read and write in that notation, drill them on how quickly they could read and write in that notation, but never play for them any actual music.

The common negative perceptions of mathematics extend well beyond this anecdote, and its ramifications involve things far more important than my own simmering inner rage. To get a sense of the larger picture, it's helpful to touch briefly on the work of Jo Boaler, a professor of mathematics education at Stanford University. Boaler has spent a lot of time thinking about effective teaching habits, and has studied how teaching habits can affect students' perception of mathematics down the road.

One of her studies compared two schools with similar demographic makeups, but in two different socioeconomic regions.[1] Boaler gave the schools pseudonyms: Phoenix Park and Amber Hill. The students at Phoenix Park tended to come from working class families, while Amber Hill's student body was more affluent. Although we sometimes think of affluent communities being more progressive in adopting new teaching practices, in this case the opposite was true. Amber Hill relied on a more traditional, lecture-based style of instruction and pedagogy, while Phoenix Park took on a more open-ended, project-based approach (see Table 9.1).

At Phoenix Park, the picture that Boaler paints of the math classroom is about as far removed as one can get from what we think of when we create a mental image of the "typical" math class:

> The classrooms at Phoenix Park did look chaotic. The project-based approach meant a lot less order and control than in traditional approaches. Instead of teaching procedures that students would practice, the teachers gave the students projects to work on that *needed* mathematical methods. From the beginning of year 8 (when students started at the school) to three quarters of the way through year 10, the students worked on open-ended projects in every lesson ... the students were taught in mixed-ability groups, and projects usually lasted for about three weeks (Boaler 2008, p. 69, emphasis original).

As a consequence of this more collaborative environment, students had a much healthier relationship with mathematics and a much better understanding of its purpose and its power:

> Not surprisingly, the Phoenix Park students came to view mathematical methods as flexible problem-solving tools. When I interviewed Lindsey in the second year of the school, she described the maths approach: "Well, if you find a rule or method, you try and adapt it to other things" (Boaler 2008, p. 74).

Another student, Simon, described math class in the following way: "You're able to explore. There's not many limits and that's what's interesting" (Boaler 2008, p. 74).

At Amber Hill, on the other hand, the classroom was much more conventional. Teachers lectured at the front of the board, students listened, then teachers assigned exercises and students did their best to solve them. Compared to the interviews with the Park Hill students, kids from Amber Hill described mathematics in a much less inspiring way. The most depressing quote came from a student named Louise, who said, "In maths you have to remember. In other subjects you can think about it" (Boaler 2008, p. 76).

Juxtapose that against "ideas are more important than math." These two sentiments were expressed by students in different communities, separated by thousands of miles. And yet, the similarity is striking. Why do I bring this up? Because for all the fun I've had writing about games from a mathematical perspective (and for all the fun I hope you've had reading), there's a critique that could reasonably be made. Namely, why should we care about this mathematical navel-gazing into video games? Don't we have more pressing educational issues in this country that need to be addressed? In an increasingly technological world, shouldn't we be more focused on improving student ability in mathematics, rather than looking at esoteric examples of mathematics that arise if you look at certain games from the right angle (no pun intended)?

It's certainly a question I've asked myself. Ultimately, though, I believe there's significant value in the types of analysis we've done here.[2] Moreover, I believe that the types of questions and ideas we've explored here could have a meaningful effect on students who, like Louise or that nameless Coloradan, incorrectly believe that *mathematical thinking* is a contradiction in terms.

9.2 WHY GAMES?

There are three reasons why I believe that turning a mathematical eye toward games, and video games in particular, is more than just a fun idea.

1 Games Are Relevant

Video games are ubiquitous among students. According to a 2008 report from the Pew Internet Project,[3] a whopping 97% of teenagers play video games (99% for boys compared to 94% for girls). Half of teenagers played video games as recently as "yesterday." And these aren't necessarily casual gamers, either: 75% of surveyed teens reported playing at least once a week, and 31% reported playing at least once a *day*.[4]

Teens aren't the only ones playing games, either. Every year, the Entertainment Software Association publishes a report on the current state of affairs in the video game industry. Their 2016 report, titled "Essential Facts about the Computer and Video Game Industry," contains the following interesting statistics on the video game audience:

- 63% of U.S. households have at least one person who plays video games three or more hours per week,
- The average game player is thirty-five years old, and 44% of game players are thirty-six or over,

- The proportion of adult (18+) women who play video games is nearly twice as large as the proportion of boys (under 18) who play video games (31% vs. 17%),
- The average number of years a gamer has been playing video games is thirteen.

Video games are a multibillion-dollar industry, even with the rise of relatively inexpensive mobile games. More people play games than don't, and almost everyone under the age of eighteen plays. Why, then, are we trying to get kids fired up about math by asking them questions about farmers who want to maximize the areas enclosed by fences, or about when two trains will cross paths? Most attempts to make math relevant to students seem trapped in the twentieth century (if not earlier).

And to be fair, maybe it was easier to make math relevant one hundered years ago. Certainly, word problems today don't seem that different than word problems from the turn of the last century. Old mathematics exams sometimes make the rounds on the Internet, as part of a larger "look how much stupider we are than our ancestors were" narrative. For example, here is a list of questions from an alleged 1895 eighth-grade final exam for a school in Salina, Kansas:[5]

1. Name and define the Fundamental Rules of Arithmetic.
2. A wagon box is 2 ft. deep, 10 feet long, and 3 ft. wide. How many bushels of wheat will it hold?
3. If a load of wheat weighs 3942 lbs., what is it worth at 50 cts. per bu., deducting 1050 lbs. for tare?
4. District No. 33 has a valuation of $35,000. What is the necessary levy to carry on a school seven months at $50 per month, and have $104 for incidentals?
5. Find cost of 6720 lbs. of coal at $6.00 per ton.
6. Find the interest of $512.60 for 8 months and 18 days at 7 percent.
7. What is the cost of 40 boards 12 inches wide and 16 ft. long at $20.00 per in?
8. Find bank discount on $300 for 90 days (no grace) at 10 percent.

9. What is the cost of a square farm at $15 per acre, the distance around which is 640 rods?
10. Write a Bank Check, a Promissory Note, and a Receipt.

Of these ten questions, only one of them (the first) is an abstract math question. The other nine are all *word problems*, and while they may not seem especially relevant now, they were probably much more relevant to the average eighth-grader in nineteenth-century Kansas. (For one thing, they don't even tell you the volume of a bushel in question two—it must have been common knowledge!) I'd also argue that analogues of many of these problems exist in modern textbooks, though they're much less relevant to a typical twenty-first century eighth-grader.

But even if these questions weren't all that relevant even to students in 1895, there's a critical difference between students then and students now: competition for time. People today, students included, are inundated with media options. Decades of television, music, and even games are now easily accessible and can be consumed on devices small enough to fit in our pockets. Because of this, it's possible that relevancy simply matters more than it did in the 1800s. If students don't think that what they're learning is relevant, it's harder to convince them that it's worth spending time on. This is especially true when there are so many engaging things competing for their time and attention.

I'm not saying that math needs to be as engaging as a video game all the time, or even most of the time. Indeed, a significant part of mastering mathematical concepts comes from skills practice, which can often feel dry and not that relevant. But people are willing to perform boring tasks all the time: athletes stick to workout regimens even though they may not be the most intellectually stimulating, and musicians practice for hours to keep their fundamentals fresh. The problem with mathematics is that we too often focus on the skills without showing students why the skills matter.

To put it simply, we often fail to make math relevant. By showing students connections between mathematics and the things they spend so much of their time on anyway, we can close the gap between what we want them to learn and what they find interesting. With so many people playing video games, there must be some low-hanging fruit here. This book is simply my attempt to pick some of it while it's ripe.

2 Games Have Pedagogical Value

In addition to being big business, games are also fodder for a fair amount of research, including research in education. Playing games can be an intensely engaging experience; a lot of very smart people are currently trying to figure out whether it's possible to use video games, or interactive media in general, to engage students as intensely in education. After all, what parent hasn't lamented the fact that his child can't sit down for fifteen minutes to do her math homework but gleefully spends an afternoon mashing buttons in front of a screen?

As you've seen, this isn't a book about math video games; it's a book about math *in* video games. Nevertheless, if you want to get some sense of the current landscape, one book I highly recommend is Keith Devlin's *Mathematics Education for a New Era: Video Games as a Medium for Learning* (2011).

In his book, Devlin makes the case for video games as an important component of mathematics education in the twenty-first century. He views games not as a way to replace resources like the teacher or even the textbook but as a powerful complement to these existing resources. Devlin focuses on games to help students master *everyday* mathematics, including an understanding of basic arithmetic, proportion, estimation, and so on. In essence, he advocates for using games to help master middle school topics that students will encounter before they come to algebra.

In fact, when it comes to algebra and the mathematics beyond it, Devlin is skeptical about the efficacy of video games in helping students achieve mastery. One reason is that everyday mathematics seeps into our world in concrete ways. Once we get to algebra and beyond, though, concrete contexts need to be stripped out, and symbolic abstraction must, at some point, reign supreme. After all, one of the reasons why mathematics is powerful is because of its general applicability; seemingly disparate questions, once stripped of their context, may be equivalent mathematically.

While this abstraction is necessary and important, Devlin argues that it's not easy to make a compelling video game around such abstraction without it feeling like the mathematics is simply being unnaturally grafted onto the game. More specifically, he argues that because the underlying educational philosophy of video games is "situated learning, they do not lend themselves naturally to teaching abstract, symbolic mathematics, such as algebra" (Devlin 2011, p. 153).[6]

All is not lost, however. When it comes to more advanced mathematics, Devlin does see potentially fruitful ground at the intersection of math and video games. But rather than being a part of the game itself, Devlin recommends designing games that will inspire mathematical thinking within the broader social community that surrounds the game. James Paul Gee, author of *What Video Games Have To Teach Us about Learning and Literacy*, calls this larger collection of social interactions the "metagame" or the "big-G game" and describes it as "all the sorts of social interactions inside or outside a game that the game inspires, encourages, or enhances."[7]

As we've seen here, many games, even ones not explicitly designed as educational video games, can inspire some serious mathematical thinking at the metagame level. Occasionally this type of thinking may help the player develop strategies to become better at the game; indeed, if you look at social communities online, you'll see that discussion of strategy makes up a significant amount of the conversation. Other times (as in the case of our discussion of *Draw Something*), mathematical thinking can help us analyze the mechanics of a game in order to better understand its strengths and weaknesses. In either case, if the game is well designed, there's the potential for interesting mathematics to emerge, regardless of whether or not the game is explicitly marketed as an educational game.

It's one thing to find interesting mathematics in games; it's another to use games effectively in the classroom in order to motivate the development of mathematics. As to how games (whether educational or not) could be used to motivate the need for mathematics at the level of the metagame, Devlin suggests using games as a way to *introduce* concepts. He describes one example in which a game designer could contextualize linear equations in terms of some game mechanic:

> So instead of solving the symbolic expression $3x + 5 = 20$, the player has to determine how many girders of length 3 units it takes to build a simple bridge across a chasm of width 20 units starting from a ledge that stretches out 5 units. To be sure, such an easy

> example can be solved quickly by inspection or trial and error, but it is not difficult to generate much harder versions where trial and error would be either unfeasible in the game context or else would take a very long time.
>
> Does experience at solving such puzzles prepare the groundwork for a student to subsequently learn the corresponding symbolic algebra? Probably not. But a teacher could take that puzzle as a starting point for a lesson on the algebra behind it, and build on the students' experience and motivation in order to teach them that algebra (Devlin 2011, p. 166).

It's a sentiment shared by Kurt Squire, author of *Video Games and Learning: Teaching and Participatory Culture in the Digital Age* (2011). In his book, Squire writes about a social studies project he helped develop using the game *Civilization*, a game in which players select an ancient civilization and try to guide it successfully through thousands of years of world history. The course focused on a question from Jared Diamond's book *Guns, Germs, and Steel* (1999): Why did Europeans colonize America instead of the reverse? Based on his experience, Squire (like Devlin) advocates for using games as introductions:

> Gaming may be most effective as a *leading activity* for academic practices. Playing *Civilization* in after-school settings led students to develop an affiliation for history that fueled game play, and each became expert in overlapping areas of *game play* and in *history*... [playing *Civilization*] increased their interest in social studies and in turn steered them into more academically valued practices such as reading books or watching documentaries (Squire 2011, p. 177).

Of course, starting every lesson with a game isn't a silver bullet: as Devlin rightly points out, there's still the issue of transferring knowledge from a specific context to a more general one. He writes: "It requires a huge cognitive leap to go from algebraic thinking in specific contexts to mastery of the associated symbolic descriptions and reasoning" (Devlin 2011, p. 166). But this is a universal challenge in mathematics education, and at the very least, using games to introduce mathematical ideas shouldn't make the problem of transfer *worse*. And in fact, if different games can be shown to have similar underlying mathematical principles, this may help students abstract their thinking.

There's one final point I'd like to make in favor of the pedagogical value of games in the classroom, and the mathematics classroom in particular. Mastery of content is certainly an important goal of the classroom, and the one that gets a lot of attention when we talk about how to improve education in this country. But there's another important goal of mathematics education that gets comparatively less attention: helping students to *think* mathematically. Keith Devlin describes it as follows: "Mathematics education, when it is successful, is only partially about people learning how to *do* mathematics; it is also about helping them adopt a particular identity—that of *being* at least a mathematically able person, and possibly even a mathematician" (Devlin 2011, p. 166).

In the National Research Council's book *Adding It Up: Helping Children Learn Mathematics*, a related notion, which they call "productive disposition," is identified as one of five interwoven strands required for mathematical proficiency. They define productive disposition as "a habitual inclination to see mathematics as sensible, useful, and worthwhile, coupled with a belief in diligence and one's own efficacy" (Kilpatrick et al. 2001, p. 116).

Even for students who have mastered the common procedures they learn in the classroom, many lack the ability to think creatively when presented with a new problem and lack confidence in their ability to tease known structures out of unfamiliar contexts. But these are essential skills in a variety of twenty-first century careers; it's no wonder that Devlin claims "a productive disposition is the *single most important outcome to aim for in K-12 mathematics education*" (Devlin 2011, p. 163, emphasis original).

This is relevant because mathematics and video games are two examples of what James Paul Gee calls *semiotic domains*. He defines a semiotic domain as "any set of practices that recruits one or more modalities (e.g., oral or written language, images, equations, symbols, sounds, gestures, graphs, artifacts, etc.) to communicate distinctive types of meanings." However, semiotic domains are probably best understood via some of the examples he provides: "cellular biology, postmodern literary criticism, first-person-shooter video games, high-fashion advertisements, Roman Catholic theology, modernist painting, midwifery, rap music, wine connoisseurship" (Gee 2007, p. 19)—these are but a few of the countless examples of semiotic domains.

An analogue of "productive disposition" could be described for any semiotic domain, and indeed the development of a productive disposition in one domain can have spillover effects into related domains. Gee describes this phenomenon using the example of his grandson, whom Gee watched playing the Nintendo game *Pikmin*. After describing his six-year-old grandson's eagerness to solve problems within the game, Gee writes:

> [T]he identity *Pikmin* recruits relates rather well to the sort of identity a learner is called on to assume in the best active learning in schools and other sites. Such learning—just like *Pikmin*—encourages exploration, hypothesis testing, risk taking, persistence past failure, and seeing "mistakes" as new opportunities for progress and learning (Gee 2007, p. 37).

In other words, properly implemented, the habits of mind that a player develops when playing a game can mirror the habits of mind we'd like students to develop in the math classroom. So not only should we look to games for inspiration when it comes to content, we should look to them for inspiration in developing a productive disposition, as well.

Of course, video games aren't the only means to encourage the development of a productive disposition. But with 97% of students already playing games, it seems wasteful to not consider this medium seriously.

3 Games Have a Precedent

Even though the game industry is massive and games are played by a majority of Americans, they are still commonly critiqued as being a waste of time. But as Gee points out, this critique on its own is fairly simplistic:

> [I]f we are concerned with whether something is worth learning or not, whether it is a waste of time or not—video games or anything else—we should start with questions like the following: What semiotic domain is being entered through this learning? Is it a valuable domain or not? In what sense? Is the learner

> learning simply to understand ("read") in parts of domain or also to participate more fully in the domain by learning to produce ("write") meanings in the domain? And we need to keep in mind that in the modern world, there are a great many more potentially important semiotic domains than just those that show up in typical schools (Gee 2007, p. 23).

Gee also rightly points out that whether a semiotic domain is valuable or not also depends on its relationship to other semiotic domains. And indeed, when it comes to games more broadly, there's a rich history of their intersection with mathematics, going back hundreds of years.

In fact, the origins of probability can be traced to the study of simple games. Sixteenth-century mathematician Gerolamo Cardano published one of the first treatises on probability, titled *Liber de Ludo Aleae* (*The Book on Games of Chance*). It provides one of the first written definitions of what we now call mathematical probability. However, the book also has its fair share of errors; for instance, Cardano attributed "luck" to a supernatural force he dubbed the "authority of the Prince."[8]

Also, unfortunately for Cardano, his book was published posthumously in 1663, almost one hundred years after he had written it. More importantly, this publication came nine years after a series of letters between mathematicians Blaise Pascal and Pierre Fermat in 1654. The birth of modern probability theory is widely credited to these two, because through these letters they solved a long-standing problem in mathematics now known as the *problem of points*.

The problem of points, like so many questions we've considered here, arises within the context of a game. In its simplest formulation, it goes something like this: suppose you and I decide to play a game that we each have a 50% chance of winning. The game could be as simple as flipping a coin, or it could be something like a round of darts, a round of bowling, or a round of *Street Fighter II*, provided that we're evenly matched. We also decide that the winner of this competition will be the first person to win some agreed-upon number of rounds of the game: it could be first-to-3, first-to-10, or first-to-n for any whole number n we choose. To make things interesting, we each throw some money into a combined pot. The amount doesn't matter—it could be $5 each or $500 each—all that matters is that we contribute the same amount.

Simple enough. But now, suppose our game is interrupted before we're able to declare a winner. For instance, in a first-to-3 series, maybe you've won two rounds to my one, when something comes up and one of us has to leave. The problem of the points asks: *given the score at the moment the game is interrupted, how should the pot be divided fairly?*

This is a problem that dates back to at least 1494, where it appears in the textbook *Summa de arithmetica, geometria, proportioni et proportionalita* by Luca Pacioli. However, the solution presented there is incorrect, for it suggests that the pot should be split in proportion to the number of rounds each person has won. So, in the case where you have two "points"—that is, you've won two rounds to my one—Pacioli's solution would be for you to take twice as much of the pot as I take. This may seem reasonable in a first-to-3 series, but it seems decidedly off in, say, a first-to-100 series, in which case at two points to one, each of us would still be nearly equally likely to win.

Other attempts were made to solve this problem, though none were successful until the middle of the seventeenth century, when Blaise Pascal learned of the problem through Chevalier de Méré, himself a gambler for whom this question may have had some practical application. Pascal wrote to Fermat about the problem; through their correspondence, not only did they solve the problem, but in so doing they also developed basic ideas in combinatorics, as well as ideas like expected value, which are now fundamental in probability theory.[9]

The solution is actually quite nice, and I'd be remiss if I didn't show you at least a piece of it. One key insight into their solution is that the number of points each person has doesn't matter: it's the number of points each person needs *in order to win* which should determine how the pot is split. For example, if we're playing a first-to-3 game and you're up 2 points to my 1, you need only one more win to clinch the victory, while I need 2. So the game would be over in at most two more rounds. The probabilities are broken out in Table 9.2.

Based on the analysis in Table 9.2, we see that you have a 3/4 = 75% chance of winning given your one-point lead, and I have a 1/4 = 25% chance of winning. In other words, you're three times as likely as I to win, so a fair division of the pot would have you take three dollars for every dollar I take.

TABLE 9.2. Steps in the Problem of Points

Round 1 Outcome (Probability)	*Round 2 Outcome (Probability)*	*Outcome of Match*	*Total Probability*
You Win (1/2)	N/A	You Win 3-1	$\frac{1}{2}$
I Win (1/2)	You Win (1/2)	You Win 3-2	$\frac{1}{2} \times \frac{1}{2} = \frac{1}{4}$
I Win (1/2)	I Win (1/2)	I Win 3-2	$\frac{1}{2} \times \frac{1}{2} = \frac{1}{4}$

TABLE 9.3. Steps in Longer Game

Round 1	*Round 2*	*Round 3*	*Round 4*	*Outcome*	*Probability*
You Win (1/2)	You Win (1/2)	N/A	N/A	You Win 4-1	$\frac{1}{4}$
You Win (1/2)	I Win (1/2)	You Win (1/2)	N/A	You Win 4-2	$\frac{1}{8}$
You Win (1/2)	I Win (1/2)	I Win (1/2)	You Win (1/2)	You Win 4-3	$\frac{1}{16}$
You Win (1/2)	I Win (1/2)	I Win (1/2)	I Win (1/2)	I Win 4-3	$\frac{1}{16}$
I Win (1/2)	You Win (1/2)	You Win (1/2)	N/A	You Win 4-2	$\frac{1}{8}$
I Win (1/2)	You Win (1/2)	I Win (1/2)	You Win (1/2)	You Win 4-3	$\frac{1}{16}$
I Win (1/2)	You Win (1/2)	I Win (1/2)	I Win (1/2)	I Win 4-3	$\frac{1}{16}$
I Win (1/2)	I Win (1/2)	You Win (1/2)	You Win (1/2)	You Win 4-3	$\frac{1}{16}$
I Win (1/2)	I Win (1/2)	You Win (1/2)	I Win (1/2)	I Win 4-3	$\frac{1}{16}$
I Win (1/2)	I Win (1/2)	I Win (1/2)	N/A	I Win 4-3	$\frac{1}{8}$

Let's do one more example. Suppose now that you're still ahead two points to my one, but now we're playing a first-to-4 match. In this case, you need two points to win and I need three. At most, there will be four more rounds of play. The possibilities are broken out in Table 9.3.

In this case, you'll win with a probability of 11/16, slightly less than the 75% from before. Conversely, I'll win with a probability of 5/16, slightly more than before. So, if our game is interrupted, we should split the pot in a ratio of 11:5; that is, you take $11 for every $5 I take.

Unfortunately for our current method of solving the problem, the number of cases to consider grows quite rapidly as you increase the number of rounds. Hence the need for a combinatorial approach, which Pascal and Fermat developed. Using modern notation, they found that if one player is r points away from a win, and another is s points away, then the pot should be split in the ratio of

$$\sum_{k=0}^{s-1} \binom{r+s-1}{k} \text{ to } \sum_{k=s}^{r+s-1} \binom{r+s-1}{k}.$$

(Here $\binom{n}{k} = \frac{n \times n-1 \times \ldots \times n-k+1}{k \times k-1 \times \ldots \times 2 \times 1}$ is called the binomial coefficient of n and k.) This agrees with the cases we've already considered: when $s = 2$ and $r = 1$, the first term is $\binom{2}{0} + \binom{2}{1} = 3$, while the second term is $\binom{2}{2} = 1$, for a ratio of 3 to 1. Similarly, when $s = 3$ and $r = 2$, the first term is $\binom{4}{0} + \binom{4}{1} + \binom{4}{2} = 11$, and the second term is $\binom{4}{3} + \binom{4}{4} = 5$, for a ratio of 11 to 5.[10]

But I digress. The point here is simply that probability theory, one of the modern cornerstones of mathematics, has grown from considering problems that arose naturally within the context of games. The problem of points may be doable by undergraduates in mathematics nowadays, but at the time Pascal and Fermat were working on it, it was genuinely new mathematics.

The physical world has provided fodder for interesting mathematical questions for centuries. But game worlds, while made by people, also provide ample opportunity for mathematical thinking. In some sense, this shouldn't be terribly surprising. After all, good games frequently involve strategic thinking and the balancing of tradeoffs. So when we're looking for interesting mathematics, it's worth looking at the games people are playing. Not because we want to pander to the audience, but because, as history has already shown, there's beautiful math to be found.

9.3 WHAT NEXT?

Today, millions of people have powerful gaming machines in their pockets and on-hand twenty-four hours a day. With that easy access, there's a renewed push into so-called "educational games." But there's rich mathematical thinking to be gleaned from games in general, not just ones that bill themselves as educational. And mathematical conversations structured around popular games are more likely to be successful than those built around games-in-name-only. The questions we've considered here haven't come from "educational games." Instead, as we've seen, interesting mathematical questions are often a by-product of good design.

Of course, in an ideal world there would be a plethora of games designed from the ground up to be both engaging and mathematically rich. There are a few examples of these types of games already; *DragonBox*, developed by WeWantToKnow, is a good example of a game that helps introduce students to algebraic concepts, and *Wuzzit Trouble* is a game that invites mathematical thinking, developed by BrainQuake, a company cofounded by Keith Devlin.

Some countries are taking things even further and building schools around games and principles of game design. Scandinavia seems particularly interested in this approach. For instance, Nordahl Grieg Upper Secondary school in Bergen, Norway, a school that opened in 2010, weaves games into a variety of its classes. *Portal* 2 is used in physics. *Skyrim* is used to explore ideas of nineteenth-century Norwegian romantic nationalism. *The Last of Us*, a popular 2013 game about the aftermath of an apocalyptic event, is studied the way that a piece of literature might be studied in a more traditional classroom. In a 2014 article on the school written for KQED *MindShift*, teachers emphasize that these games are taken seriously:[11]

> Tobias Staaby uses video games to teach units ranging from ethics to narrative and cultural history. "I wanted to use video games as something more than chocolate covered broccoli," he said. "It's important that video games are regarded as useful and engaging learning tools in their own right." To that end, he uses popular commercial games that would not outwardly seem

> suitable for the classroom. . . . [G]ame-based learning seems to be a misnomer, as the learning is not based on games, but enhanced by them. Commercial games are repurposed and modified to support curricular goals, as opposed to driving them.

In other words, while we wait for the potential of educational games to catch up to the reality of the current gaming landscape, popular games can already be used as valuable educational resources, when placed in the hands of educators who are passionate about using games to support learning. For math specifically, I hope to have convinced you that rich, interesting mathematics is present in many of the games we already play. And with so many of us playing games, there are plenty of shared experiences we could be mining to help introduce mathematical concepts and develop mathematical thinking.

Long story short: there's a lot of mathematics under the surface of many of our most popular video games, if you're willing to look. Beautiful mathematics doesn't discriminate. And neither should we.

Notes

Introduction

1. Like this one: $\zeta(s) = 2^s \pi^{s-1} \sin\left(\frac{\pi s}{2}\right) \Gamma(1-s) \zeta(1-s)$. There's more where that came from.

Chapter 1. Let's Get Physical

1. If you'd like more examples of questions aimed at uncovering physics misconceptions, an excellent resource is Epstein (2009).
2. If we take air resistance and wind into account, even the picture on the left isn't totally accurate, because the diver's horizontal velocity may decrease while in the air.
3. Let's ignore the physical hurdles we'd need to overcome in order for a platform to hover several feet above the ground and freely float through space.
4. Mint-condition versions of the original *Castlevania* game can sometimes be found on eBay listed for thousands of dollars.
5. For one Mario-based example, at Midwood High School in Brooklyn, New York, students of physics teacher Glenn Elert analyzed seven different Mario games to study how the acceleration due to gravity has changed in each game. They discovered that, generally speaking, the acceleration due to gravity represented in Mario games has become closer to the true value over time, though it's still much higher than in real life. See Lefky and Gindin (2007) for more details.
6. Media Molecule developed the first two games in the main series, *LittleBigPlanet* in 2008 and *LittleBigPlanet 2* in 2011. The third game in the series, *LittleBigPlanet 3*, was developed by Sumo Digital and released in 2014.
7. The nine-million-level milestone was announced on the *LittleBigPlanet*'s Facebook page: https://www.facebook.com/littlebigplanet/posts/10152307038246831.
8. See, for instance, https://www.youtube.com/watch?v=I-2-vdNItoI for an example of the types of contraptions people can build with the tools Media Molecule has provided.
9. Speaking as one of many students who had to build a Rube Goldberg device in physics class, I can certainly see the merit in starting with a virtual machine first.

10. If you're interested in seeing any of the levels created for competition, search for "Contraption Challenges" on YouTube. If you have a copy of *LittleBigPlanet* (1, 2, or 3), you can search for the levels in the game with the text "CC##," with ## replaced by the number of the challenge you'd like to see. The flying machine challenge was number 4, and the Rube Goldberg challenge was number 6, but there have been plenty of other cool challenges as well!
11. This foundation is perhaps best known for its annual MacArthur Fellowship awards (more colloquially known as "genius awards"), given to a handful of people each year who "show exceptional merit and promise for continued and enhanced creative work." The award comes with a significant no-strings-attached cash prize. Since the award's inception in 1981, twenty-seven mathematicians have been named MacArthur Fellows.
12. If you have a copy of *LittleBigPlanet* (1, 2, or 3) and are curious about the levels created for this competition, the winning entries are titled "Aeon Quest: Abduction," "Discovery Pier: A Whole New Spin on Science and Engineering," "LittleBigChemistryLab," "Sackboys and the Mysterious Proof," and "Stem Cell Sackboy."
13. We need to be a bit careful about the terminology here. The momentum of an object that travels through a portal is conserved relative to the portal. For example, if you put two portals next to each other on the same wall, it would be possible for you to play catch with yourself. Throw a ball through one portal and it will emerge from the other, with its velocity essentially reversed. In this sense, the momentum of the ball is *not* conserved. However, from the portal's perspective, nothing has changed; the ball simply passed through it without any change in momentum. So, the statement "portals conserve momentum" should be understood in this latter context (indeed, the game itself informs the player that portals conserve momentum, though it doesn't worry too much about the details).
14. The law of conservation of energy states that in a closed system, energy can be neither created nor destroyed. It can change forms, however; for example, when an airplane lands on a runway, its kinetic energy is transferred to heat energy due to air resistance and friction between the wheels and the runway. The total energy of this system, however, remains the same.
15. Here we're choosing our reference frame so that the floor has a height of zero. This is more a matter of convenience than anything else—you're free to set the zero height to be somewhere else; the general argument will remain the same.
16. Such perpetual motion machines must violate the laws of thermodynamics. The impossibility of such a machine is wellknown in the scientific community, and is also recognized by the United States Patent and Trademark

Office, which refuses to grant any patent submitted for a perpetual motion machine that is not accompanied by a working model. This extra condition is not a requirement for any other kind of device.

17. To read about one high school teacher's experience using *Portal* in the classroom, check out www.physicswithportals.com, written by Cameron Pittman.
18. GPS technology, by the way, needs to incorporate the effects of special relativity in order to determine your location with such precision. Relativity may seem like heady stuff, but there are many practical applications!
19. The idea of using a game to explore this thought experiment, however, precedes the MIT Game Lab. In 2007, researchers C. M. Savage, A. C. Searle, and L. McCalman released an application called "Real Time Relativity" with a similar goal.
20. The game is available for free from the MIT Game Lab website.
21. In fact, the Michelson–Morley experiment performed near the end of the nineteenth century strongly suggested that the speed of light is a constant, and much of the mathematical groundwork needed for special relativity was completed by physicists George FitzGerald and Hendrik Lorentz. It was Einstein, however, who brought everything together and showed how the effects of relativity, though counterintuitive, were nevertheless both natural and necessary.
22. For a behind-the-scenes view on game development, the documentary *Indie Game: The Movie* features Phil Fish, designer of *Fez*, during the final stages of its production.
23. Math and computer science aren't the only subjects that have been infiltrated by *Minecraft*. According to "Teaching in the Age of Minecraft," a 2015 article in *The Atlantic*, "History teachers make Minecraft dioramas, English teachers have kids act out Shakespeare plays in a model of the Globe Theater, and art teachers let students re-create famous works of art in the game" (Ossola 2015).

    Sometimes these projects combine together in innovative ways. In *The Game Believes in You: How Digital Play Can Make Our Kids Smarter*, author Greg Toppo describes a musical production at Virginia Tech University that was set inside of *Minecraft*. Students built sets inside the game. Singers positioned themselves next to the screen while other students controlled in-game avatars from backstage. Of the performance, Toppo wrote: "This isn't the same old thing done more efficiently or, heaven forbid, the same old stupid thing done more efficiently. These aren't better flash cards or gamified toothbrushes with points attached. This is cutting-edge technology applied to something totally new and strange and beautiful. It offers kids the chance to create something no one has ever seen, something keyed to their passions but with an eye toward broadening them. They start with one foot in a familiar world and end up somewhere new and different.

In this case, of course, they're stepping from the digital world back into ours. Our gamers became students" (Toppo 2015, p. 217).

24. For a nice introduction on the history of computer science and the fundamental importance of logic gates, check out Petzold (2000).
25. Have you ever seen what happens when sulfuric acid reacts with sugar? Search for it on YouTube. You won't be sorry.

Chapter 2. Repeat Offenders

1. For more details on *Family Feud*'s ratings, see Kissell (2015).
2. These numbers are pulled from Morris (2010). Among other things, this guide lists all questions, answers, and point values for the 1993 Super Nintendo *Family Feud* game.
3. This model may not be entirely accurate. For example, there are probably safeguards in place to ensure that you won't see the same question twice in a row. Factoring in these additional rules would make the model more complicated, though, and wouldn't affect the larger takeaway in any significant way.
4. If you like a little more rigor with your probability theory, here's a way to derive the formula using the language of random variables. Let's treat the general case. Suppose you're drawing with replacement from a pool of N items. Let X be the number of draws needed to obtain the first repeat. Then the probability there are no repeats within the first k draws is equal to $P(X > k)$. By the definition of conditional probability, we have

$$\begin{aligned} P(X > k) &= P(X \neq k \cap X > k-1) \\ &= P(X \neq k \mid X > k-1) \times P(X > k-1) \\ &= \frac{N-(k-1)}{N} \times P(X > k-1), \end{aligned}$$

since if the first repeat doesn't happen in the first $k - 1$ draws, there are $N - (k - 1)$ possibilities for the kth draw that will not result in a duplicate. Note that this recursive formula relates $P(X > k)$ to $P(X > k - 1)$; since we know that $P(X > 1) = 1$, this gives us our formula by applying the above equality $k - 1$ times. If you prefer, you can also show that this gives the desired formula by using induction.

5. This value of 27 questions is unchanged if we replace the 500 questions assumed in the discussion with any of the actual numbers from the *Family Feud* game, which all hover around the number 500. Don't believe me? Feel free to run the numbers yourself!
6. To get this formula, we again need to use a bit of probability theory. Instead of looking at $R(k, N)$, the number of repeats after k draws from a pool of N objects, let's look at $k - R(k, N)$, the expected number of *unique objects* seen

after k draws. For each number j between 1 and N, let $X_{j,k}$ be the random variable equal to 1 if the jth object makes an appearance in the first k draws, and 0 otherwise. Then the number of unique objects seen after k draws is equal to

$$X_{1,k} + X_{2,k} + \ldots + X_{N,k}.$$

This, in turn, means that the expected number of unique objects seen is equal to the expected value of this sum. In other words,

$$\begin{aligned} k - R(k, N) &= E(X_{1,k} + X_{2,k} + \ldots + X_{N,k}) \\ &= E(X_{1,k}) + E(X_{2,k}) + \ldots + E(X_{N,k}). \end{aligned}$$

Since nothing distinguishes one object from another within the collection, we also know that

$$E(X_{1,k}) = E(X_{2,k}) = \ldots = E(X_{N,k}),$$

so that

$$k - R(k, N) = NE(X_{1,k}).$$

Lastly, what is $E(X_{1,k})$? Well, by the definition of expected value, since $X_{1,k}$ takes on only the values of 1 and 0,

$$E(X_{1,k}) = 1 \times P(X_{1,k} = 1) + 0 \times P(X_{1,k} = 0) = P(X_{1,k} = 1).$$

On the other hand, we see that this probability equals

$$1 - P(X_{1,k} = 0) = 1 - \left(1 - \frac{1}{N}\right)^k,$$

since on any given draw, the probability of not getting the first object is $1 - 1/N$, and there are k draws. Therefore

$$k - R(k, N) = N\left(1 - \left(1 - \frac{1}{N}\right)^k\right),$$

and this gives us the formula for $R(k, N)$.

7. The probability that the next round will be new can be calculated in a similar way as $R(k, N)$. Using the same notation as in our previous derivation, the next round will be new precisely when

$$(X_{1,k+1} = 1 \cap X_{1,k} = 0) \cup \ldots \cup (X_{N,k+1} = 1 \cap X_{N,k} = 0).$$

Since these are mutually exclusive events, the probability that this happens is equal to

$$\begin{aligned} \sum_{i=1}^{N} P\,(X_{i,k+1} = 1 \cap X_{i,k} = 0) &= \sum_{i=1}^{N} P\,(X_{i,k+1} = 1 \mid X_{i,k} = 0)\, P\,(X_{i,k} = 0) \\ &= \sum_{i=1}^{N} \frac{1}{N}\left(1 - \frac{1}{N}\right)^k = \left(1 - \frac{1}{N}\right)^k. \end{aligned}$$

8. See Yenigun (2013).
9. Sale figure taken from Chen and Wortham (2012).
10. Even this number may be an underestimate, because for a time, *Draw Something* was a highly social game. People shared pictures of their favorite drawings on Facebook, and entire websites were created for the sole purpose of displaying particularly good drawings. Therefore, the number of words people saw may have been even higher. Of course, we're also assuming that each word in a triplet is equally likely to be selected; this is unlikely, but won't affect the analysis too much.
11. See, for instance, Sliwinksi (2013).
12. Notice that when the total number of word triplets increases by a factor of five, after a month of use the expected percentage of repeats decreases by a factor slightly less than five. In fact, if you increase the number of questions by a factor of c, it's true that you'll decrease the number of expected repeats for a fixed k by a factor which is on the order of $1/c$. Here's a proof: consider the ratio

$$\frac{R(k, cN)}{R(k, N)} = \frac{k - cN\left(1 - \left(1 - \frac{1}{cN}\right)^k\right)}{k - N\left(1 - \left(1 - \frac{1}{N}\right)^k\right)}.$$

For fixed k and N, we can use the approximation $\left(1 - \frac{1}{cN}\right)^k \approx e^{-k/cN}$, meaning that

$$\frac{R(k, cN)}{R(k, N)} \approx \frac{k - cN\left(1 - e^{-k/cN}\right)}{k - N\left(1 - e^{-k/N}\right)}.$$

Since k and N are fixed, the denominator is just some constant. For the numerator, if we use the Taylor series expansion for e^x, we find that it equals

$$k - cN\left(1 - \left(1 - \frac{k}{cN} + \frac{1}{2!}\left(\frac{k}{cN}\right)^2 - \frac{1}{3!}\left(\frac{k}{cN}\right)^3 + \ldots\right)\right)$$

$$= k - cN\left(\frac{k}{cN} - \frac{1}{2!}\left(\frac{k}{cN}\right)^2 + \frac{1}{3!}\left(\frac{k}{cN}\right)^3 + \ldots\right)$$

$$= \frac{1}{c}\left(\frac{1}{2!}\frac{k^2}{N} - \frac{1}{3!}\frac{k^3}{cN^2} + \ldots\right),$$

which is on the order of $1/c$, again because N and k are fixed.

Bonus questions: How does the ratio above vary with n and/or N? What happens as n grows? As N grows?

13. If you'd like to be a little more precise, here's how you can describe these shifting probabilities in the language of probability theory. Suppose you have N questions and draw one per round for k rounds. For $1 \leq i \leq N$ and $1 \leq j \leq k$, let $X_{i,j}$ equal 1 if the question in round j is the ith question, and 0 otherwise. Also, let $S_{i,m}$ denote the number of times that question i has appeared after the first m rounds; in other words, $S_{i,m} = \sum_{j=1}^{m} X_{i,j}$. Finally, suppose our weight factor is p; in other words, the question that is drawn in the first round should be p times as likely to be drawn the second round, compared to the other questions. In this case, the probability that the question that is drawn in round $m + 1$ is the ith question is $P(X_{i,m+1} = 1)$, and

$$P\left(X_{i,m+1} = 1\right) = \frac{p^{S_{i,m}}}{\sum_{n=1}^{N} p^{S_{n,m}}}.$$

Note that when $p = 1$, this reduces to $1/N$, which is exactly what we get in the case where you draw questions with replacement.

14. A similar effect emerges if you look at the probability that the next draw will be new. Looking at this example doesn't add much more to the conversation, so I'll leave it for you to explore if you're curious.

15. Put more precisely, if you've seen m questions, the number of rounds you need to play before seeing your next new question follows a geometric distribution with probability $p_{m,N} = \frac{N-m}{N}$, and therefore its expected value is $\frac{1}{p_{m,N}} = \frac{N}{N-m}$.

16. This problem is more commonly known as the *coupon collector's problem*, and has a relatively well-known solution. As N grows, the expected number of rounds is well approximated by the expression

$$N \ln N + \gamma N + 1/2,$$

where $\gamma \approx 0.577$ is the Euler–Mascheroni constant.

17. In the language of calculus, $1 + x$ is the first-degree Taylor polynomial for e^x. Since the Taylor series for e^x converges to e^x at every point, this explains why $e^x \approx 1 + x$ near 0.

18. Well known is, of course, a relative term. In case you haven't seen this formula, here's one way to convince yourself why it's true: write the sum in two rows, first moving from the smallest number to the largest, and then vice versa:

$$1 + 2 + \ldots + (k-2) + (k-1)$$

$$(k-1) + (k-2) + \ldots + 2 + 1.$$

If we add these two rows together, we see that twice the sum is equal to $k+k+\ldots+k$. This gives us $(k-1)$ copies of k, so twice the sum equals $k(k-1)$, and therefore the sum must equal $k(k-1)/2$.

It's said that Gauss discovered this trick as a child when a particularly sadistic teacher asked his students to find the sum of the first one hundred numbers. If you want to one-up a young mathematical prodigy, can you quickly find the sum of the first one hundred squares? Cubes?

Chapter 3. Get Out the Voting System

1. This isn't an argument for the plurality system in the case of major elections between two candidates, such as the election of the president. Even though the United States is essentially a two-party system, there do exist third-party candidates who make it onto the ballot. Moreover, even for the major-party candidates, there are still preliminary elections that help decide who inside of the two largest parties will get the nod for the general election. In fact, the principle known as Duverger's law asserts that the two-party system is a result of, not a justification for, the plurality voting system.
2. In spite of the somewhat complicated rules, instant runoff voting is one of the most common alternative voting systems in wide adoption. In California, it is used for the election of most citywide offices in San Francisco and Oakland and is used for some offices in neighboring Berkeley. Since 2010, it's also been used to select the Best Picture winner at the Academy Awards.
3. This assumes, of course, that the voters are acting sincerely and aren't casting ballots that don't align with their true feelings.
4. One commonly cited practical argument against instant runoff voting is that it isn't additive. This means that ballots can't be tabulated at a district level, with subtotals sent along to a central location. Instead, all of the ballots must be collected and centralized first, and then they must be counted altogether. Here's an example to illustrate why: suppose that of the 100 ballots cast in our Microsoft/Sony/Nintendo election, 56 came from people in one town, and 44 from people in another. Suppose further that those first 56 votes broke down like this:

TABLE 3.2. Ballots from the First Town in Our Hypothetical Scenario

1st place	*2nd place*	*3rd place*	*Number of Votes*
Microsoft	**Nintendo**	Sony	30
Sony	**Nintendo**	Microsoft	26

Given the totals, this means that the second set of votes looked like this:

TABLE 3.3. Ballots from the Second Town in Our Hypothetical Scenario

1st place	2nd place	3rd place	Number of Votes
Microsoft	**Nintendo**	Sony	15
Nintendo	Sony	Microsoft	15
Sony	**Nintendo**	Microsoft	4
Sony	Microsoft	**Nintendo**	10

According to IRV, in both towns Microsoft would have won (it has a majority in the first town and a majority in the second town after Sony is eliminated in the first round). However, even though Microsoft is the winner in both towns individually, Sony is the winner overall, since when the votes are tallied together, Nintendo drops out first. In other words, with IRV, determining winners among subsets of the voting population isn't useful.

5. It should also be said that Arrow's theorem also makes some (intuitive and reasonable) assumptions on voters' preferences. Namely, voters should be capable of ordering all candidates into a list of preferences, and those preferences should be transitive; that is, if you prefer candidate A to candidate B, and candidate B to candidate C, you must prefer candidate A to candidate C.
6. This is the gist of score voting, though there are some smaller details to consider. For instance, to ensure that a write-in candidate who appears on only one ballot doesn't win with a perfect score of five stars, there are sometimes rules about how many scores a candidate has to receive in order to be eligible for the win.
7. Because approval voting lets you vote for as many candidates as you like, some people contend that it violates "one person, one vote." However, the confusion here is purely semantic. A "vote" in the context of approval voting is simply a ballot with a list of candidates who have been marked as approved or not. Everyone still has the same voting power.
8. To learn more about the Center for Election Science (CES) arguments in favor of approval voting, see https://www.electology.org/approval-voting. Full disclosure: I served as a contributing member for CES from 2011 through early 2014.
9. To see the Center for Election Science's comparison of IRV and approval voting, visit https://www.electology.org/approval-voting-vs-irv.
10. For a more detailed definition, see Smith's article "Bayesian regret for dummies" at http://scorevoting.net/BayRegDum.html.
11. See the article titled "Assassin's Creed 4: Black Flag—'Best' Mission Revealed By Ubisoft" on nowgamer.com.

12. Interestingly, you can still give other *players* negative ratings, based on their comments and reviews of other levels.
13. The argument in this section is drawn from Miller (2009).
14. One review of the game on Time.com is titled "No, You Don't Have to Play Flappy Bird" (Peckham 2014).
15. For more details on these claims, see Silverwood (2014).
16. See Hamburger (2014).
17. The Wilson score confidence interval isn't the only game in town. To learn about some alternatives, as well as see some criticisms of this approach, check out https://news.ycombinator.com/item?id=1218951.
18. A proof of why the variance is equal to $p(1 - p)/n$ can be found in many probability textbooks, but for the sake of completeness, here's a quick explanation. The best way to think about it is to view each play of a level as a Bernoulli trial with probability p of success. In other words, P is the probability that someone will like the given level; each Bernoulli trial will equal 1 with probability P and will equal 0 otherwise. With n plays, we have n Bernoulli trials; let's call them $X_1, X_2, \ldots, X_n$. The sum of these trials, S_n denotes the number of likes, and S_n/n denotes the proportion of likes. Then what we want to show is that the variance in S_n/n, written $\text{Var}(S_n/n)$, is equal to $p(1 - p)/n$.

    The variance of a random variable X is equal to $\text{Var}(X) = E(X^2) - E(X)^2$, where $E(Y)$ denotes the expected value of Y. From this formula, one can show that $\text{Var}(cX) = c^2\text{Var}(X)$ for any constant c, and if X and Y are independent, then $\text{Var}(X + Y) = \text{Var}(X) + \text{Var}(Y)$. Assuming the plays of our *LittleBigPlanet* level are independent, this means that

$$\text{Var}\left(\frac{S_n}{n}\right) = \frac{1}{n^2}\text{Var}(S_n) = \frac{1}{n^2}\text{Var}\left(\sum_{i=1}^{n} X_i\right) = \frac{1}{n^2}\sum_{i=1}^{n}\text{Var}(X_i).$$

    Meanwhile, what is the variance of any one of the Bernoulli trials? Take the first one, for instance. X_1 has value 1 with probability p and value 0 with probability $(1 - p)$. Therefore, X_1^2 also has value 1 with probability p and value 0 with probability $(1 - p)$, since $1^2 = 1$ and $0^2 = 0$. This means that

$$\text{Var}(X_1) = E(X_1^2) - E(X_1)^2 = p - p^2 = p(1 - p).$$

    Since every Bernoulli trial has a variance of $p(1 - p)$, and there are n trials, the sum of the variances is equal to $np(1 - p)$, and the variance for S_n/n equals $np(1 - p)/n^2 = p(1 - p)/n$, which is just what we wanted.
19. Of course, there's nothing special about the choice of a 95% probability. We could easily adjust this number up or down, provided that we also adjust the multiple of the standard error (1.96) accordingly.

Chapter 4. Knowing the Score

1. If you're curious, you can sample some of the games here: http://www.ferryhalim.com/orisinal/. Go on, try them out. I'll meet you back here in a few hours.
2. If you're curious and have an Apple device, the mobile version of the game can be yours for $0.99.
3. One wrinkle here is that if you make a particularly smooth landing onto a car, you get a "smooth landing" bonus instead of a "nice landing" bonus. The smooth landing bonus is twice as large as the nice landing bonus, which complicates our later analysis just a bit. But if you're interested in making a more exhaustive analysis, be my guest!
4. This, incidentally, provides an algebraic proof of the fact that two consecutive triangular numbers always sum to a perfect square.
5. It turns out that the number of cases you need to consider isn't quite the number of partitions, but the fact that $p(n)$ grows so quickly should still make us skeptical that this is a reasonable approach. Some partitions we can eliminate right off the bat: for instance, if we know c and S, and we assume that $n_1 \geq n_2 \geq \ldots \geq n_{k-1} \geq 1$, then we can eliminate any partition with $n_1 \geq \sqrt{S}$. On the other hand, for each partition we want to consider, we have to make a choice about which whole number in the partition we want to equal n_k, which will *increase* the number of cases we have to check.
6. These numbers were obtained with the use of technology. If you really want to explore partitions and see how quickly they grow, try writing a little computer program which, for given values of S and c, will calculate how many possible solutions there are to the set of equations

$$S = n_1^2 + n_2^2 + \cdots + n_{k-1}^2 + \frac{n_k(n_k - 1)}{2},$$

$$c = n_1 + n_2 + \ldots + n_k.$$

7. When $k = 50$, the unique solution is given by $n_1 = 34$, $n_2 = 7$, $n_3 = 4$, $n_5 = n_6 = 2$, and $n_i = 1$ otherwise. When $k = 52$, the unique solution is given by $n_1 = 34$, $n_2 = 8$, $n_3 = n_{52} = 2$, and $n_i = 1$ otherwise. When $k = 53$, the unique solution is given by $n_1 = 34$, $n_2 = 8$, $n_3 = 2$, and $n_i = 1$ otherwise.
8. One can modify Jacobi's formula for the sum of two squares to eliminate this sort of double counting. This elimination of redundancy reduces the number of solutions by a factor of roughly 1/8; in many cases, such as the example here, the factor is exactly 1/8.
9. Jacobi's results on sums of two and four squares are closely tied to the theory of modular forms. These types of results are beautiful but require a fair amount of prerequisite knowledge. For a treatment that considers the cases mentioned here, check out Hardy and Wright (1979).

10. If you're curious, here are the data on the playthrough used as the example throughout this section: $n_1 = 10$, $n_2 = 3$, $n_3 = 11$, $n_4 = 22$, $n_5 = 4$, $n_6 = 5$, $n_7 = 19$, $n_8 = 11$, and $n_9 = 9$.
11. For an interesting article on the imitation *Threes!* inspired in the wake of its release, check out Vanhemert (2014).
12. Of course, there are other ways one could tinker with the model as well. For instance, it's highly unlikely that the final board will have $c_i = 8$ for any values of i; it's much more likely that the values of each c_i will be less than 8, especially for higher-value cards. Incorporating this fact into the model could yield some interesting results.

    One could also go the other way and try to generalize the mathematics that comes out of *Threes!* For example, what if there are k point-scoring cards instead of twelve? What if the board size is $j \times j$ instead of 4×4? How does varying these parameters affect the number of ways you can obtain a given score?
13. They asserted as much when they published an infographic on *Threes!* when the game was released for the Android operating system. See, for example, http://cdn2.sbnation.com/assets/4127853/threes-infographic_960.png.
14. In fact, it's always the case that for any score S, there is a unique board configuration that minimizes $c_1 + c_2 + \ldots + c_{12}$. Can you convince yourself why this is the case, and determine every other configuration from this minimal one?
15. Most of this section is adapted from an April 2016 post on my blog, *Math Goes Pop!*
16. See Hernandez (2016) for an interesting glimpse into the world of speedrunning.
17. The Hamming distance is an actual distance in the mathematical sense. The distance between two words is always nonnegative and is equal to zero precisely when the words are the same. Moreover, the Hamming distance is symmetric (the distance from word_1 to word_2 is the same as the distance from word_2 to word_1), and it satisfies the triangle inequality (the distance from word_1 to word_3 is no larger than the distance from word_1 to word_2 plus the distance from word_2 to word_3).
18. Hint: You might want to explore what's known as the *Levenshtein distance*, a sort of generalization of the Hamming distance.
19. If you search for *edit distance* or *Levenshtein distance* on your favorite web browser, you'll find a lot of quick information on applications. One paper worth exploring if you're interested in a biological application is Dinu and Sgarro (2006).
20. There are a number of relevant resources here for the reader who wants to learn more, though the results are fairly technical. Here's an overview of some of them. For an exploration of the limits to running speeds in different

animals, see Denny (2008). An analysis of Olympic records (both limits to the records and which records have the most room for improvement) can be found in Einmahl and Magnus (2012). Unfortunately, many of their predictions were later falsified as records that they anticipated would be difficult to improve were later broken. For a discussion of this, see Hilbe (2009). A different analysis focusing only on the 100-m race for men and women can be found in Einmahl and Smeets (2011). Lastly, for a more general resource on extreme-value theory, see Ferguson (1996).

21. This high-score data is taken from http://donkeykongblog.blogspot.com/2011/12/donkey-kong-world-record-history.html.
22. See, for example, Good (2016).
23. The R^2 value for the logistic model given here is roughly 0.93. This is roughly the same as the R^2 value you get if you ignore the 1982 record.
24. See http://www.classicarcadegaming.com/forums/index.php/topic,881.msg8309.html.

Chapter 5. The Thrill of the Chase

1. By the end of its life cycle, the Nintendo 64 had a worldwide library of 387 games. By comparison, Sony's first PlayStation console, a direct competitor to the N64, had a library of more than 2,400 games.
2. For the purists, yes, I mean Koopa shells.
3. Within this context, one of the roots is always negative. Can you explain why?
4. This is not entirely true. Players have some degree of control over v_K. At higher difficulties, all the karts move faster, which increases the value of v_K. Also, for a given difficulty, different characters have different speed characteristics. For example, Toad accelerates quickly but has a relatively low top speed. Wario, on the other hand, accelerates more slowly but has a higher top speed.
5. Here's a bonus exercise: suppose your opponent's distance to you is k times the width of the track. What is the angle in terms of k? Can you expand your expression into an infinite series involving k? What sorts of heuristics do the first few terms of this series give you?
6. A limit calculation shows that

$$\lim_{m\to\infty} \theta_m = \cos^{-1}\left(\frac{v_K}{v_S}\right).$$

Notice that the larger v_S is compared to v_K, the larger this limiting angle. This makes sense: if your shell is much, much faster than your opponent, it can afford to bounce up and down several times along the way.

If it's only a bit slower than v_K, though, too much vertical bouncing may slow its horizontal motion enough that your opponent can escape.

7. When m increases by 1, the distance the opponent travels increases by $v_K(t_{m+1} - t_m)$. With a little work, it's possible to show that

$$\lim_{m\to\infty}(t_{m+1} - t_m) = \frac{w}{\sqrt{v_S^2 - v_K^2}}.$$

Therefore, each additional bounce off the wall increases the distance your opponent travels by

$$\frac{v_K w}{\sqrt{v_S^2 - v_K^2}} = \frac{w}{\sqrt{(v_S/v_K)^2 - 1}}.$$

Again, there are a couple of ways to check your intuition that the above formula makes sense. For instance, it makes sense that the *wider* the road is, the *larger* the above distance is. Similarly, the *faster* your shell is relative to your opponent's kart, the *shorter* the above distance is.

8. Another common approach involves saturating the course with shells. If you fire every green shell you come across, eventually one is bound to hit something, right? While such a strategy may increase the likelihood of a hit, you're just as likely to hit an opponent as you are yourself.
9. Of course, we already know that nothing can travel faster than the speed of light.
10. Fans of *Missile Command* may have a quibble here: namely, we have assumed that the missiles from each battery have the same speed. In fact, the missiles from the middle battery travel faster than the ones on the outsides, which adds another layer of complexity to the game. However, our analysis focuses on only one missile/counter-missile pair, so designating a single missile speed for the player doesn't hurt our analysis. The interested reader can explore what the math tells us as v_P decreases or increases.
11. It's worth confirming this algebraically as well, without using the geometric fact that S encloses the origin.
12. Can you find a general formula for the area of S?
13. For a detailed description of ghost behavior in *Pac-Man*, check out Pittman (2009).
14. There are some other details of their movement; for example, occasionally all ghosts will reverse their directions.
15. The ghost movement algorithms are all based on an emulated version of the game available at webpacman.com.
16. For all the juicy mathematical details, See Foderaro, Swingler, and Ferrari (2012).

17. According to Twin Galaxies, the organization responsible for maintaining and verifying the accuracy of arcade high-score records, the highest score achieved in *Ms. Pac-Man* is 933,580. This record was achieved by Abdner Ashman on April 6, 2006.
18. The derivation here closely follows the one presented in Nahin (2012).
19. Unlike with green shells, the case $k = 1$ is actually interesting when it comes to red shells. Try to work out the details on your own, if you're curious.
20. It would be nice to consider generalizations where you fire the shell *before* your opponent begins to round the corner of the course, but it's not possible to come up with a closed-form solution for the red shell's path using calculus in this case. If you're curious, try it out for yourself and see where you get stuck. For extra credit, you can use a computer to model this situation and see what kind of trajectories you get!

Chapter 6. Gaming Complexity

1. Tetrominoes are one example of a more general class of shape called a polyomino. Polyominoes are formed by arranging an arbitrary number of linked squares. As any *Tetris* player knows, there are seven distinct tetrominoes (assuming you aren't allowed to flip pieces, i.e., the J and the L count as distinct pieces). By these rules, can you figure out how many pentominoes (5 squares) there are? What about hexominoes (6 squares)? What about the number of polynomials composed of n squares?
2. For example, see Holmes et al. (2009).
3. Don't believe computers can calculate Bacon numbers quickly? You can go to The Oracle of Bacon (oracleofbacon.org) to see how quickly it's done, using any pair of actors that your heart desires.
4. To get a sense for how infeasible this approach would be, let's do some quick estimation. The Internet Movie Database currently has more than 5 million film professionals in its system. Even if we suppose that only 10% of those people are actors (as opposed to directors, screenwriters, producers, etc.), that's at least 500,000 people we can use to build paths to Kevin Bacon.

   Suppose now that we want to search for a path between Kevin Bacon and, say, Samuel L. Jackson, that goes through 1,000 films. That means we need to find 999 actors, with no repeats. If we simply compile every list of 999 actors, and then see if we can form a path between Bacon and Jackson through that list, we'll have around 500,000 choices for the first actor, 499,999 for the second, 499,998 for the third, and so on. In total, this means the number of lists is around $500{,}000 \times 499{,}999 \times \ldots \times 499{,}903 \times 499{,}902 \approx 1.56 \times 10^{564}$.

   Even on a computer that could check trillions of lists a second, this number is impossibly, almost incomprehensibly large. A computer working since the beginning of time would be able to check only the most insignificant fraction of the total number of possible lists. In other words, this strategy is a nonstarter.

5. The movies that provide the links in Table 6.1 are, in order, *Django Unchained* (2012), *The Amazing Spider-Man 2* (2014), *The Help* (2011), *Get on Up* (2014), *Night at the Museum: Battle of the Smithsonian* (2009), *Reality Bites* (1994), *Big Trouble* (2002), *(500) Days of Summer* (2009), *Inception* (2010), and *Super* (2010).
6. Somewhat more precisely, the set **P** consists of problems with yes-or-no answers that can be solved in polynomial time with respect to the length of an input. In other words, if we ask the problem about a string of length n, and the problem can be solved in at most cn^k steps for some c and k positive, then the problem is in **P**.
7. Somewhat more precisely, **NP** stands for nondeterministic polynomial time. These are problems for which solutions can be verified in polynomial time.
8. In fact, there are physical implications of the **P** vs. **NP** problem. In a 2014 paper, physicist Arkady Bolotin at Ben-Gurion University in Israel published a paper showing why macroscopic quantum objects can't exist if $\mathbf{P} \neq \mathbf{NP}$. This would explain why phenomena that are common in quantum mechanics, such as quantum superposition, don't occur at scales we can observe in our everyday experience. For more details, see Bolotin (2014).
9. For a general overview, check out Fortnow (2013). More technical references can be found there.
10. The algorithm wouldn't be able to solve a board if some guessing is required.
11. For examples of these levels in action, head to YouTube and search for "Impossible Levels."
12. See, for example, https://www.youtube.com/watch?v=c6IfLRh5cMY.
13. More specifically, the results of Greg Aloupis and his colleagues apply to *Super Mario Bros. 1, 3, the Lost Levels* (NES) and *Super Mario World* (SNES) from the *Super Mario* series; *Donkey Kong Country 1–3* (SNES) from the *Donkey Kong Country* series; *Metroid* (NES) and *Super Metroid* (SNES) from the *Metroid* series, *The Legend of Zelda* (NES), and *The Legend of Zelda: A Link to the Past* (SNES) from the Zelda series.
14. Though in the most recent entry, *inFAMOUS: Second Son*, the player has no health bar. Instead, blast shards are used to strengthen the character's superpowers.
15. The root of this fetch quest can actually be traced to a somewhat rushed development cycle. The original version of *The Wind Waker* was meant to have more content, but due to time constraints some of it had to be scrapped. This particular fetch quest helped lengthen the game, at the expense of the story's pacing. For more on this, see Welsh (2013).

16. An application called Concorde will solve (some) TSP problems for you: an online version can be found at http://neos.mcs.anl.gov/neos/solvers/co:concorde/TSP.html. For the example presented here, the input is a list of pairs of distances between each of the fragments, as well as distances to the headquarters:

	HQ	1	2	3	4	5	6	7	8	9	10	11
HQ	0											
1	1240	0										
2	1474	789	0									
3	1127	972	497	0								
4	360	1032	1131	767	0							
5	551	1139	1065	628	240	0						
6	835	1481	1230	735	612	387	0					
7	774	1604	1422	930	642	466	208	0				
8	414	1609	1701	1271	600	644	724	569	0			
9	1149	2196	2038	1543	1179	1063	808	622	758	0		
10	1788	2768	2489	1996	1785	1631	1300	1165	1399	641	0	
11	1613	2769	2741	2260	1738	1695	1540	1334	1144	783	941	0

The numbers in each cell tell us distances between different objects. For example, the headquarters is at a distance of 0 from itself, a distance of 1240 from the first data fragment, a distance of 1474 from the second data fragment, and so on. The distances are with respect to an arbitrary scale, so the numbers themselves don't matter as much as the ratios between them. For instance, the distance between the first fragment and seventh fragment is more than twice as large as the distance between the first and second fragments, since $1604 \div 789 \approx 2.03$.

17. Why doesn't the problem get simpler if you forgo the requirement that you finish where you start? Well, having an open path is equivalent to having a closed path where the starting/ending point is at a distance of 0 from every other point. Of course, this isn't something we can easily visualize, ingrained as we are in our Euclidean notions of distance, but mathematically there's no difference. Indeed, for the open-path version of the current example, the only difference with the input data is that the first column is replaced by zeros, indicating that the start/end point is some imagined location that has a distance of zero from every data fragment:

	HQ	1	2	3	4	5	6	7	8	9	10	11
HQ	0											
1	0	0										
2	0	789	0									
3	0	972	497	0								
4	0	1032	1131	767	0							
5	0	1139	1065	628	240	0						
6	0	1481	1230	735	612	387	0					
7	0	1604	1422	930	642	466	208	0				
8	0	1609	1701	1271	600	644	724	569	0			
9	0	2196	2038	1543	1179	1063	808	622	758	0		
10	0	2768	2489	1996	1785	1631	1300	1165	1399	641	0	
11	0	2769	2741	2260	1738	1695	1540	1334	1144	783	941	0

18. In this case, the addition of the fast-travel spots has the effect of decreasing some of the distances in the previous list:

	HQ	1	2	3	4	5	6	7	8	9	10	11
HQ	0											
1	0	0										
2	0	789	0									
3	0	972	497	0								
4	0	1032	1069	740	0							
5	0	1139	1065	628	240	0						
6	0	1481	1230	735	612	387	0					
7	0	1604	1422	930	642	466	208	0				
8	0	1609	1124	794	600	644	724	569	0			
9	0	1560	1029	700	680	871	808	622	734	0		
10	0	2115	1585	1255	1236	1427	1300	1165	1290	641	0	
11	0	1846	1316	986	967	1157	1367	1334	1021	783	941	0

19. If you're savvy, you may have noticed that I said there were 100 data fragments in all, and yet 11 + 77 is not 100. The discrepancy is caused by a third area, called Cappadocia, which contains the remaining twelve data fragments.

20. Curious about TSP with moving targets? Check out Li, Yang, and Kang (2006), along with the references therein.
21. To date, the largest TSP solved with an exact solution consists of a network of 85,900 nodes. Details can be found in Applegate et al. (2011). The traveling salesman problem has also been solved for the most efficient way to visit all 24,978 towns in Sweden. For details, visit http://www.math.uwaterloo.ca/tsp/sweden/.
22. Because finding exact solutions can prove so daunting, there are times when approximate solutions are used to find paths that are quite short, though there's no guarantee they'll be the shortest possible. This can speed up calculation time, if finding the absolute shortest path isn't as important as finding one that's relatively short.

Chapter 7. The Friendship Realm

1. Spoiler alert: the magnitude of the buddy scores actually has no effect on gameplay; it's only the relative rankings of the scores that matter. In fact, regardless of the size of the buddy score, your best buddy, along with some of your other friends, will betray you in the final moments of the game. In order to escape, you're forced to gun them down.
2. This table is taken from http://www.uesp.net/wiki/Skyrim:Disposition.
3. This terminology was introduced in Gottman (2002).
4. The evaluation of this sum is a commonly encountered problem in calculus, if not sooner. The easiest way to verify the formula is to assign a variable to the value of the sum, say S. In other words,

$$S = \sum_{i=1}^{t} r^{t-i} = r^{t-1} + r^{t-2} + \cdots + r + 1.$$

If we multiply S by r, we get the related sum

$$rS = r^t + r^{t-1} + \cdots + r^2 + r.$$

If we arrange S and rS in a particular way, we see that there are a number of common terms, so subtracting them yields a lot of cancellation:

$$\begin{aligned} S &= r^{t-1} + r^{t-2} + \cdots + r^2 + r + 1, \\ -rS &= -r^t - r^{t-1} - r^{t-2} + \cdots - r^2 - r, \\ \Longrightarrow S - rS &= -r^t + 1. \end{aligned}$$

In other words, $S - rS = (1-r)S = 1 - r^t$, so $S = \frac{1-r^t}{1-r}$.

5. Assuming the amount of effort you can put into a relationship on any day is bounded by some absolute maximum, it's possible to invoke the limit

argument even earlier: it's true that

$$\lim_{t\to\infty} f(t) = \lim_{t\to\infty} a \sum_{i=1}^{t} r^{t-i} \times c(i).$$

For a challenge, you can try to prove why this statement is true. (Hint: It's not necessarily obvious why the limit on the right-hand side even exists, so this is not as straightforward as it might seem.)

6. In all of this work, we assumed that the amount of effort spent every day was the same, which doesn't seem realistic. How would the model change if we assumed that the friends met only once a week but spent the same amount of effort on the friendship during their weekly meetings? Can you find a range of values for the long-term friendship score if the two people put in an effort between c_1 and c_2 every day?
7. Just as we assumed that friendship scores were mutual, we'll assume their feelings are modeled by the same utility and disutility functions, to keep things simple.
8. For example, if D is a differentiable function of c, we're good to go.
9. There are several assumptions on the functions U and D that we didn't make explicit since our discussion was more general. To be more precise, his conclusions are valid when U is at least twice differentiable with U' positive and U'' negative, and U' tending to 0 as the friendship score tends to infinity (so that the marginal increase in utility with friendship score decreases). Similarly, he assumes that D is at least twice differentiable with $D'' > 0$, $D' = 0$ at a single point (corresponding to a unique minimum) and D' tending to infinity as the amount of effort tends toward infinity. In other words, after reaching its minimum value, the marginal increase in disutility grows without bound as effort increases.
10. If you know a little calculus, you can verify directly that the two solutions given really do satisfy the required differential equations. If you know a little more calculus, you can verify that these are the *only* solutions to the differential equations.
11. In this more general setting, it's still possible to find closed-form solutions to the differential equations. If you're able, try to show that the solutions take the following form:

$$M(t) = e^{\frac{x(a+d)}{2}} \left[\left(\frac{(a-d)M_0 + 2bP_0}{\sqrt{D}} \right) \sinh\left(\frac{x\sqrt{D}}{2} \right) + M_0 \cosh\left(\frac{x\sqrt{D}}{2} \right) \right],$$

$$P(t) = e^{\frac{x(a+d)}{2}} \left[\left(\frac{(d-a)P_0 + 2cM_0}{\sqrt{D}} \right) \sinh\left(\frac{x\sqrt{D}}{2} \right) + P_0 \cosh\left(\frac{x\sqrt{D}}{2} \right) \right],$$

where $D = (a - d)^2 + 4bc$. Be sure to convince yourself why the equations make sense regardless of the sign of D.

12. *Systems theory* has names for these qualitatively different types of phenomena: reinforcing and balancing feedback loops. Reinforcing feedback loops are ones in which an initial change in some quantity causes even greater change in that quantity. Exponential growth is an example of a reinforcing feedback loop. Balancing feedback loops are ones in which an initial change yields an effect that tends to diminish said change. Therefore, balancing feedback loops tend to promote stability.
13. With our earlier models, taking $M_0 = P_0 = 0$ would have resulted in $M(t) = P(t) = 0$ for all t, which, I think you'll agree, isn't so interesting.
14. In fact, all things being equal, whenever Peach's intrinsic appeal increases, the percent increase in Mario's long-term feelings will be higher than the percent increase in Peach's long-term feelings. The opposite is true if Mario's intrinsic appeal increases. You can prove this directly from the formulas for M_{stable} and P_{stable} (you'll also need the inequality $ae > bd$). In other words, self-improvement certainly benefits you, but according to the model it benefits your partner even more. Or, as Rinaldi puts it, "there is a touch of altruism in a woman (man) who tries to improve her (his) appeal."
15. For a discussion of some of these more advanced models, see Bielczyk et al. (2012).
16. Or at least, as Einstein is alleged to have said.

Chapter 8. Order in Chaos

1. To simplify the notation somewhat, in this section we'll denote a function only by its letter (e.g., M), and not its letter and its argument (e.g., $M(t)$).
2. Note that sensitivity to initial conditions is a necessary condition for chaos, but according to most definitions, is not sufficient. For example, all exponential functions show sensitivity to initial conditions, but they aren't chaotic. However, sensitivity to initial conditions is the simplest component of chaos to see and understand, and it's the one we'll focus on here. Other aspects of chaos are beyond the scope of our present discussion.
3. The proof that $\sqrt{2}$ is irrational is ancient: the Pythagoreans knew it, for instance. Here's a quick proof: if $\sqrt{2}$ is rational, we can write it as p/q where p and q are integers. If p and q have any common factors, we can divide them out, and so it does us no harm to assume that the fraction p/q is in lowest terms (i.e., that p and q have no common factors). Since $p/q = \sqrt{2}$, it must be true that $p^2 = 2q^2$; since the right-hand side is even, the left-hand side must be too, which means that p must be even. But if p is even, we can write it as $p = 2r$ for some integer r. In other words,

$$2q^2 = p^2 = (2r)^2 = 4r^2.$$

Dividing both sides by 2 again, we see that $q = 2r$. This, in turn, means that q is even. But if q is even, then both p and q had a common factor, contradicting our earlier assumption that p/q was in lowest terms! Because we've discovered a contradiction, the only logical conclusion is that our initial assumption—namely, that $\sqrt{2}$ is rational—must be false.

4. This assumes that the shell is a point that has no width, perfectly reasonable mathematically, though within the game world it's less realistic.
5. To understand *why* irrational slopes lead to this space-filling behavior, it's best to visualize the board not as a square in the plane but as a doughnut (or, if you want to be fancy, a *torus*) in three-dimensional space. If you're curious about why squares and doughnuts have anything to do with one another, see Freiberger (2014b).
6. Here's one (informal) way you could enumerate the rational numbers. First, draw a sequence of number lines; for each successive row, increase the denominator by 1. Then, starting at 0/1, follow the arrows to pass through the diagram, skipping any numbers that are already on your list:

$$\begin{array}{cccccccccccccc}
\cdots & -\frac{3}{1} & -\frac{2}{1} & \leftarrow & -\frac{1}{1} & & \frac{0}{1} & \rightarrow & \frac{1}{1} & & \frac{2}{1} & \rightarrow & \frac{3}{1} & \cdots \\
 & \uparrow & \downarrow & & \uparrow & & & & \downarrow & & \uparrow & & \downarrow & \\
\cdots & -\frac{3}{2} & -\cancel{\frac{2}{2}} & & -\frac{1}{2} & \leftarrow & \cancel{\frac{0}{2}} & \leftarrow & \frac{1}{2} & & \cancel{\frac{2}{2}} & & \frac{3}{2} & \cdots \\
 & \uparrow & \downarrow & & & & & & & & \uparrow & & \downarrow & \\
\cdots & -\cancel{\frac{3}{3}} & -\frac{2}{3} & \rightarrow & -\frac{1}{3} & \rightarrow & \cancel{\frac{0}{3}} & \rightarrow & \frac{1}{3} & \rightarrow & \frac{2}{3} & & \cancel{\frac{3}{3}} & \cdots \\
 & \vdots & \vdots & & \vdots & & \vdots & & \vdots & & \vdots & & \vdots &
\end{array}$$

Using this approach, the first few rational numbers are 0, 1, 1/2, 1/3, −1/3, −1/2, −1, ...

7. If you think carefully about decimal expansions, you may object to enumerating numbers in this way, because not all numbers have a single decimal expansion. For example, $1 = 1.000000\ \ldots = 0.999999\ \ldots$. This is a valid objection, but it can be accounted for without significantly altering the structure of the argument.
8. Otherwise, their union—the real numbers—would also be countable. Can you figure out a way to show that the union of two countable sets is still countable?
9. "Almost all" is a phrase which we can make mathematically precise, but to do so is beyond the scope of our present conversation. Pick up a book on real analysis if you're curious!
10. It should be noted that while almost all trajectories will be ergodic, periodic orbits are still dense within the space of all orbits. In other words, for any two initial slopes that yield an ergodic orbit, you can find a slope in between them that will be periodic.

11. For more on rational polygonal billiards geared toward a general audience, check out Freiberger (2014a). For a more technical approach, one popular resource is Masur and Tabachnikov (2002).
12. As before, we're restricting our attention to a 2D model of the Block Fort course; the actual course has ramps that let players drive onto buildings, but all walls are still straight, as in our simplified model.
13. If you're looking for more information on the mathematics between elliptical billiards, check out Bandres and Gutiérrez-Vega (2004) and Chernov and Markarian (2006).
14. See Stark (2014).
15. In case you're curious, Lorenz's equations are typically written as

$$x' = \sigma y - \sigma x,$$
$$y' = \rho x - zx - y,$$
$$z' = xy - \beta z,$$

for some parameters σ, ρ, and β. Note that while the differential equation for x is linear, the ones for y and z are not.
16. Indeed, probabilistic weather models are relatively popular textbook problems, usually in the study of something called *Markov chains*. These models usually give you two possible weather options (e.g., sunny or cloudy) and some information about future weather patterns given the current weather. For instance, you might be told that if it's sunny today, there's an 80% chance it will be sunny tomorrow and a 20% chance it will be cloudy; meanwhile, if it's cloudy today, there's a 60% chance it will be cloudy tomorrow and a 40% chance it will be sunny. One fun exercise (which you can explore on your own, if you're so inclined) is calculating long-term weather patterns: on average, what percentage of days should we expect to be sunny (or cloudy)?
17. One of the more popular modifications is called "Pure Weather" and can be found at nexusmods.com/skyrim/mods/52423/.
18. To be fair, *Donkey Kong Country Returns* wasn't the first Nintendo game to feature a Super Guide. It was actually the third, behind *New Super Mario Bros. Wii*, released in 2009, and *Super Mario Galaxy 2*, released earlier in 2010.
19. See Fletcher (2011).
20. Here's a question to think on: Is the question of deciding whether or not a level is beatable in **NP** if the game uses a Super Guide?

Chapter 9. The Value of Games

1. For more on this study and other work that she has done, see Boaler (2008).
2. And even if there isn't value from an applications standpoint, there's definitely value from a "mathematical interesting-ness" standpoint. For some people (myself included), that's more than enough.
3. Lenhart, et al. (2008).
4. Here, however, a gender disparity starts to emerge: fully 88% of boys reported playing at least one day a week, and 39% reported playing at least once a day; for girls, the numbers were 60% and 26%, respectively.
5. This list is retrieved from http://www.snopes.com/language/document/1895exam.asp. Even if these exact questions weren't featured on an exam from that time period, the exam is similar to others from that time.
6. Marc ten Bosch, creator of *Miegakure*, echoed these sentiments in Chapter 1.
7. See http://jamespaulgee.com/archdisp.php?id=69&scateg=Video+Games.
8. For more on Cardano's contributions to the early history of probability theory, check out Gorroochurn (2012).
9. For an interesting read on the correspondence between Pascal and Fermat, check out Devlin (2010).
10. Feeling curious? Here's a couple of other questions to poke at:

    1. If $s = n$ and $r = n - 1$, show that the payout ratio tends to 1:1 as n tends to infinity. In other words, a one-point advantage matters less and less if the number of rounds to win is large.
    2. We've assumed throughout that each person has an even chance of winning an individual round, but what if that isn't the case? How would the payout ratios look if one person has a probability p of winning a round?

11. For the full article, see Darvasi (2014).

Bibliography

Abbott, E. A. (2006). *Flatland: A Romance of Many Dimensions*. Oxford: Oxford University Press.

Aloupis, G., E. D. Demaine, A. Guo, and G. Viglietta (2015). "Classic *Nintendo* games are (computationally) hard." Theoretical Computer Science *586*(C), 135–60.

Applegate, D. L., R. E. Bixby, V. Chvatal, and W. J. Cook (2011). *The Traveling Salesman Problem: A Computational Study*. Princeton, NJ: Princeton University Press.

"Assassin's Creed 4: Black Flag—'Best' mission revealed by Ubisoft." (2013). *NowGamer*, https://www.nowgamer.com/assassins-creed-4-black-flag-best-mission-revealed-by-ubisoft.

Bandres, M. A., and J. C. Gutiérrez-Vega (2004). "Classical solutions for a free particle in a confocal elliptic billiard." *American Journal of Physics 72*(6), 810–17.

Bielczyk, N., M. Bodnar, and U. Foryś (2012). "Delay can stabilize: Love affairs dynamics." *Applied Mathematics and Computation 219*(8), 3923–37.

Boaler, J. (2008). *What's Math Got to Do with It?: How Parents and Teachers Can Help Children Learn to Love Their Least Favorite Subject*. New York: Penguin.

Bobin, J., J. Clement, and B. McKenzie (2009). "Murray Takes It to the Next Level." *Flight of the Conchords*. Directed By Troy Miller.

Bolotin, A. (2014). "Computational solution to quantum foundational problems." arXiv preprint *arXiv:1403.7686*.

Bozon, M. (2007). "Into the everybody votes channel". IGN.

Brams, S. and P. Fishburn (1978). "Approval voting." *American Political Science Review 72*(3), 831–47.

Breukelaar, R., E. D. Demaine, S. Hohenberger, H. J. Hoogeboom, W. A. Kosters, and D. Liben-Nowell (2004). "Tetris is hard, even to approximate." *International Journal of Computational Geometry & Applications 14*(01n02), 41–68.

Chatfield, T. (2011). *Fun Inc.: Why Gaming Will Dominate the Twenty-First Century*. New York: Open Road Media.

Chen, B. X., and J. Wortham (2012). "A game explodes and changes life overnight at a struggling start-up." *The New York Times*, March 25.

Chernov, N., and R. Markarian (2006). *Chaotic Billiards*, Vol. 127. Providence, RI: American Mathematical Society.

Darvasi, P. (2014). "Literature, ethics, physics: It's all in video games at this Norwegian school." *MindShift*. KQED.

Davis, J. (2012). "Why *Draw Something* blew up, but might fade fast." IGN.

Denny, M. W. (2008). "Limits to running speed in dogs, horses and humans." *Journal of Experimental Biology 211*(24), 3836–49.

Devlin, K. (2010). *The Unfinished Game: Pascal, Fermat, and the Seventeenth-Century Letter that Made the World Modern.* New York: Basic Books.

Devlin, K. (2011). *Mathematics Education for a New Era: Video Games as a Medium for Learning.* Boca Raton, FL: CRC Press.

Diamond, J. (1999). *Guns, Germs, and Steel: The Fates of Human Societies.* New York: W. W. Norton & Company.

Dinu, L. P., and A. Sgarro (2006). "A low-complexity distance for DNA strings." *Fundamenta Informaticae 73*(3), 361–72.

Einmahl, J. H., and J. R. Magnus (2012). "Records in athletics through extreme-value theory." *Journal of the American Statistical Association 103*(484), 1382–91.

Einmahl, J. H., and S. G. Smeets (2011). "Ultimate 100-m world records through extreme-value theory." *Statistica Neerlandica 65*(1), 32–42.

Entertainment Software Association (2016). *Essential Facts about the Computer and Video Game Industry.* essentialfacts.theesa.com.

Epstein, L. C. (2009). *Thinking Physics: Understandable Practical Reality* (3rd ed.). San Francisco: Insight Press.

Ferguson, T. S. (1996). *A Course in Large Sample Theory,* Vol. 49. Boca Raton, FL: Chapman & Hall.

Fletcher, J. C. (2011). "Retro reflects on *Donkey Kong Country Returns,* denies sequel plans." *Joystiq.*

Foderaro, G., A. Swingler, and S. Ferrari (2012). "A model-based cell decomposition approach to on-line pursuit-evasion path planning and the video game *Ms. Pac-Man.*" In *2012 IEEE Conference on Computational Intelligence and Games (CIG),* pp. 281–87. Granada: IEEE.

Fortnow, L. (2013). *The Golden Ticket: P, NP, and the Search for the Impossible.* Princeton, NJ: Princeton University Press.

Freiberger, M. (2014a). "Chaos on the billiard table." *Plus Magazine,* May 8.

Freiberger, M. (2014b). "Playing billiards on doughnuts." *Plus Magazine,* May 27.

Gee, J. P. (2007). *What Video Games Have to Teach Us about Learning and Literacy: Revised and Updated Edition.* Palgrave Macmillan and Houndmills, Basingstoke, Hampshire, U.K.

Good, O. (2016). "Donkey Kong's all-time record broken again, with a 'perfect' game." *Polygon.*

Gorroochurn, P. (2012). "Some laws and problems of classical probability and how Cardano anticipated them." *Chance 25*(4), 13–20.

Gottman, J. M. (2002). *The Mathematics of Marriage: Dynamic Nonlinear Models.* Cambridge, MA: MIT Press.

Hamburger, E. (2014). "Indie smash hit 'Flappy Bird' racks up $50k per day in ad revenue." *The Verge.*

Hardy, G. H., and S. Ramanujan (1918). "Asymptotic formulae in combinatory analysis." *Proceedings London Mathematical Society* (17), 75–115.

Hardy, G. H., and E. M. Wright (1979). *An Introduction to the Theory of Numbers.* Oxford: Oxford University Press.

Hernandez, P. (2016). "Why it's taking years to shave seconds off the world record for Super Mario Bros." *Kotaku.*

Hilbe, J. M. (2009). "Note: Modeling future record performances in athletics." *Journal of the American Statistical Association 104*(487), 1293–94.

Holmes, E. A., E. L. James, T. Coode-Bate, and C. Deeprose (2009). "Can playing the computer game 'Tetris' reduce the build-up of flashbacks for trauma? A proposal from cognitive science." *PloS ONE 4*(1), e4153.

Indie Game: The Movie (2012). Directed by Lisanne Pajot and James Swirsky. BlinkWorks Media.

Isbell, S. (2012). "Sack it to me: A new update and a lot of yays." *PlayStation.Blog.*

Jacobi, C. G. J. (1829). *Fundamonta Nova Theoriae Functionum Ellipticarum.* Königsberg, Germany.

Jurassic Park (1993). Directed by Steven Spielberg. Universal Pictures.

Kaye, R. (2000). "Minesweeper is NP-complete." *The Mathematical Intelligencer 22*(2), 9–15.

Kessler, S. (2012). "How a mobile game dominated the app store without any press." *Mashable.*

Kilpatrick, J., J. Swafford, and B. Findell, eds. (2001). *Adding It Up: Helping Children Learn Mathematics.* Washington, DC: National Academies Press.

King of Kong: A Fistful of Quarters, The (2007). Directed by Seth Gordon. Picturehouse.

Kissell, R. (2015). "Ratings: 'Family Feud' tops all of syndication for first time." *Variety.*

Ladner, R. E. (1975). "On the structure of polynomial time reducibility." *Journal of the ACM (JACM) 22*(1), 155–71.

Lefky, A., and A. Gindin (2007). "Acceleration due to gravity: Super Mario Brothers. "In G. Elert, ed., *The Physics Factbook,* hypertextbook.com/facts/.

Lenhart, A., J. Kahne, E. Middaugh, A. R. Macgill, C. Evans, and J. Vitak (2008). "Teens, video games, and civics: Teens' gaming experiences are diverse and include significant social interaction and civic engagement." *Pew Internet & American Life Project.* http://pewinternet.org/files/old-media/Files/Reports/2008/PIP_Teens_Games_and_Civics_Report_FINAL.pdf.pdf.

Li, C., M. Yang, and L. Kang (2006). "A new approach to solving dynamic traveling salesman problems." In *Asia-Pacific Conference on Simulated Evolution and Learning,* pp. 236–43. Berlin: Springer.

Lorenz, E. N. (1963). "Deterministic nonperiodic flow." *Journal of the Atmospheric Sciences 20*(2), 130–41.

Masur, H., and S. Tabachnikov (2002). "Rational billiards and flat structures." *Handbook of Dynamical Systems 1,* 1015–89. Elsevier, Amsterdam.

Miller, E. (2009). "How not to sort by average rating." *Evanmiller.org.*

Morris, D. (2010). Family Feud walkthrough. *Neoseeker.*

Nahin, P. J. (2012). *Chases and Escapes: The Mathematics of Pursuit and Evasion.* Princeton, NJ: Princeton University Press.

Narcisse, E. (2012). "Remember *Draw Something*? Millions of people don't like it anymore." *Kotaku.*

Nova (1997). "The Proof." PBS.

NowGamer (2013). *"Assassin's creed 4: Black flag*—'Best' mission revealed by Ubisoft." https://www.nowgamer.com/assassins-creed-4-black-flag-best-mission-revealed-byubisoft/.

Ossola, A. (2015). Teaching in the age of Minecraft. *The Atlantic*. Feb. 6.
Ottwell, G. V. (1999). "Arithmetic of Voting." Universal Workshop.
Pacioli, L. (1494). *Summa de arithmetica, geometria, proportioni et proportionalita*, Paganini, Venice.
Peckham, M. (2014). "No, you don't have to play Flappy Bird." *Time*. Feb 3.
Petzold, C. (2000). *Code: The Hidden Language of Computer Hardware and Software*. Redmond, WA: Microsoft Press.
Pittman, J. (2009). The *Pac-Man* dossier. *Gamasutra*.
Poundstone, W. (2008). *Gaming the Vote: Why Elections Aren't Fair (and What We Can Do about It)*. New York: Hill and Wang.
Rey, J.-M. (2010). "A mathematical model of sentimental dynamics accounting for marital dissolution." *PloS ONE* 5(3), e9881.
Riendeau, D. (2013). "*Donkey Kong Country Returns 3D* review: The wild side." *Polygon*.
Rinaldi, S. (1998). "Love dynamics: The case of linear couples. "*Applied Mathematics and Computation* 95(2), 181–92.
Savage, C. M., A. C. Searle, and L. McCalman (2007). "Real time relativity: Exploratory learning of special relativity." *American Journal of Physics* 75(9), 791–98.
Silverwood, M. (2014). "How in-app review mechanics pushed Flappy Bird to the top of the charts." *VentureBeat*.
Sliwinksi, A. (2013). "Zynga shutting down four games, OMGPOP site." *Joystiq*.
Smith, W. D. "Bayesian regret for dummies." *RangeVoting.org*.
Sprott, J. (2004). "Dynamical models of love." *Nonlinear Dynamics, Psychology, and Life Sciences* 8(3), 303–14.
Squire, K. (2011). *Video Games and Learning: Teaching and Participatory Culture in the Digital Age*. New York: Teachers College Press.
Stark, C. (2014). 'Mario Kart 8' review: Striking, inventive and inviting. *Mashable*, mashable.com/2014/05/15/mario-kart-8-review/#rxM4ocL+Ecqr.
Strogatz, S. H. (1988). "Love affairs and differential equations." *Mathematics Magazine* 61(1), 35.
Sullivan, C., J. Sallabank, A. Higgins, and A. Foden (2014). "It's a-me density!" *Physics Special Topics* 13(1).
Thompson, C. (2016). "The Minecraft generation." *The New York Times*. April 14.
Toppo, G. (2015). *The Game Believes in You: How Digital Play Can Make Our Kids Smarter*. New York: Macmillan.
Vanhemert, K. (2014). "Design is why 2048 sucks, and *Threes* is a masterpiece." *Wired.com*.
Weber, R. J. (1997). Comparison of voting systems. Cowles Foundation Discussion Paper No. 498, Yale University, New Haven, CT.
Welsh, O. (2013). "The Wind Waker's missing dungeons were reused in other Zelda games." *Eurogamer*.
Yenigun, S. (2013). "Calling the shots: Realistic commentary heightens video games." *NPR*.

Index